Information Sources in

Grey Literature

Fourth Edition

Guides to Information Sources

A series under the General Editorship of
Ia C. McIlwaine,
M.W. Hill and
Nancy J. Williamson

This series was known previously as 'Butterworths Guides to Information Sources'

Other titles available include:

Information Sources in Official Publications
    edited by Valerie J. Nurcombe
Information Sources in Law (Second edition)
    edited by Jules Winterton and Elizabeth M. Moys
Information Sources in the Life Sciences
    edited by H.V. Wyatt
Information Sources in Architecture and Construction (Second edition)
    edited by Valerie J. Nurcombe
Information Sources in Finance and Banking
    by Ray Lester
Information Sources in Environmental Protection
    edited by Selwyn Eagle and Judith Deschamps
Information Sources in Chemistry (Fourth edition)
    edited by R.T. Bottle and J.F.B. Rowland
Information Sources in Physics (Third edition)
    edited by Dennis F. Shaw
Information Sources in Sport and Leisure
    edited by Michele Shoebridge
Information Sources in Patents
    edited by C.P. Auger
Information Sources for the Press and Broadcast Media
    edited by Selwyn Eagle
Information Sources in Information Technology
    edited by David Haynes
Information Sources in Pharmaceuticals
    edited by W.R. Pickering
Information Sources in Metallic Materials
    edited by M.N. Pattern
Information Sources in the Earth Sciences (Second edition)
    edited by David N. Wood, Joan E. Hardy and Anthony P. Harvey
Information Sources in Cartography
    edited by C.P. Perkins and R.B. Barry
Information Sources in Polymers and Plastics
    edited by R.T. Adkins

Information Sources in
Grey Literature
Fourth Edition

C.P. Auger

London · Melbourne · Munich · New Providence, N.J.

m 011.02 AUG

British Library Cataloguing in Publication Data
A catalogue record for this title is available from the British Library

Library of Congress Cataloging-in-Publication Data
Auger, Charles P. (Charles Peter)
   Information sources in grey literature / C.P. Auger. – 4th edn.
      p. cm. – (Guides to information sources)
   Includes bibliographical references and index.
   ISBN 1–85739–194–2 (alk. paper)
   1. Technical literature.   2. Technical reports.   3. Grey literature.
I. Title.   II. Series: Guides to information sources (London, England)
T10.7.A87 1988
025.2'8–dc21   97–33378
                                                                 CIP

Published by Bowker-Saur, Maypole House, Maypole Road.
East Grinstead, West Sussex RH19 1HU, UK
Tel: +44(0)1342 330100 Fax: +44(0)1342 330191
E-mail: lis@bowker-saur.com
Internet Website: http://www.bowker-saur.com/service/
Bowker-Saur is part of REED BUSINESS INFORMATION

ISBN 1–85739–194–2

Cover design by Calverts Press
Phototypeset by Florencetype Ltd, Stoodleigh, Devon
Printed on acid-free paper
Printed and bound in Great Britain by Antony Rowe Ltd, Chippenham

Series editors' foreword

The second half of 20th century has been characterized by the recognition that our style of life depends on acquiring and using information effectively. It has always been so, but only in the information society has the extent of the dependence been recognized and the development of technologies for handling information become a priority. These modern technologies enable us to store more information, to select and process parts of the store more skilfully and transmit the product more rapidly than we would have dreamt possible only 40 years ago. Yet the irony still exists that, while we are able to do all this and are assailed from all sides by great masses of information, ensuring that one has what one needs just when one wants it is frequently just as difficult as ever. Knowledge may, as Johnson said in the well known quotation, be of two kinds, but information, in contrast, is of many kinds and most of it is, for each individual, knowable only after much patient searching.

The aim of each Guide in this series is simple. It is to reduce the time which needs to be spent on that patient searching; to recommend the best starting point and sources most likely to yield the desired information. Like all subject guides, the sources discussed have had to be selected, and the criteria for selection will be given by the individual editors and will differ from subject to subject. However, the overall objective is constant; that of providing a way into a subject to those new to the field or to identify major new or possibly unexplored sources to those already familiar with it.

The great increase in new sources of information and the overwhelming input of new information from the media, advertising, meetings and conferences, letters, internal reports, office memoranda, magazines, junk mail, electronic mail, fax, bulletin boards etc. inevitably tend to make one reluctant to add to the load on the mind and memory by consulting books and journals. Yet they, and the other traditional types of printed material, remain for many purposes the most reliable sources of information. Despite all the information that is instantly accessible via the new technologies one still has to look thongs up in databooks, monographs, journals, patent specifications, standards, reports both official and commercial, and on maps and in atlases. Permanent recording of facts, theories and opinions is still carried out primarily by publishing in printed form. Musicians still work from printed scores even though they are helped by sound recordings. Sailors still use printed charts and tide tables even though they have radar and sonar equipment.

However, thanks to computerized indexes, online and CD-ROM, searching the huge bulk of technical literature to draw up a list of references can be undertaken reasonably quickly. The result, all too often, can still be a formidably long list, of which a knowledge of the nature and structure of information sources in that field can be used to put it in order of likely value.

It is rarely necessary to consult everything that has been published on the topic of a search. When attempting to prove that an invention is genuinely novel, a complete search may seem necessary, but even then it is common to search only obvious sources and leave it to anyone wishing to oppose the grant of a patent to bear the cost of hunting for a prior disclosure in some obscure journal. Usually, much proves to be irrelevant to the particular aspect of our interest and whatever is relevant may be unsound. Some publications are sadly lacking in important detail and present broad generalizations flimsily bridged with arches of waffle. In any academic field there is a 'pecking order' of journals so that articles in one journal may be assumed to be of a higher or lower calibre than those in another. Those experienced in the field know these things. The research scientist soon learns, as it is part of his training, the degree of reliance he can place on information from co-workers elsewhere, on reports of research by new and (to him) unknown researchers on data compilations and on manufacturers of equipment. The information worker, particularly when working in a field other than his own, faces very serious problems as he tries to compile, probably from several sources, a report on which his client may base important actions. Even the librarian, faced only with recommending two or three books or journal articles, meets the same problem though less acutely.

In the Bowker-Saur Guides to Information Sources we aim to bring you the knowledge and experience of specialists in the field. Each author regularly uses the information sources and services described and any tricks of the trade that the author has learnt are passed on.

Nowdays, two major problems face those who are embarking upon research or who are in charge of collections of information of every kind. One is the increasingly specialized knowledge of the user and the concomitant ignorance of other potentially useful disciplines. The second problem is the trend towards cross-disciplinary studies. This has led to a great mixing of academic programmes – and a number of imprecisely defined fields of study. Courses are offered in Environmental Studies, Women's Studies, Communication Studies or Area Studies, and these are the forcing group for research. The editors are only too aware of the difficulties raised by those requiring information from such hybrid subject field and this approach, too, is being handled in the series alongside the traditional 'hard disciplines'.

Guides to the literature have a long and honoured history. Marion Spicer of SRIS recently drew to our attention a guide written in 1891 for engineers. No doubt there are even earlier ones. Nowadays, with the information and even the publishing fields changing quite frequently, it is necessary to update guides every few years and this we do in this present Series.

Michael Hill
Ia McIlwaine
Nancy Williamson

Preface

In literature, colour coding in the literal sense is nothing new – for years in the United Kingdom we have had yellow backs (cheap, sensational novels in yellow boards or paper covers and commonly sold on railway stations); White Papers (official documents dating from 1899 and printed on white paper); Green Papers (statements of proposed government policy dating from 1967 and intended as a basis for Parliamentary discussion, provided with green covers); Blue Books (official reports issued in blue covers); Black Books (important books bound in black such as the *Black Book of the Exchequer*); and Red Data Books (containing lists of endangered species). The latest addition to the colour range, namely grey, refers not to the physical appearance of a document but to the uncertain status of it, and gradually the term 'grey literature' is becoming more and more accepted. Nowadays most comprehensive desk dictionaries record the expression, as, for example, *The Chambers dictionary* (1993) which states: 'grey literature – material published non-commercially, eg government reports'.

The encouraging reception afforded to the third edition of this work and the requirement so soon for an updated version underline on the one hand the growing importance of grey literature as an information source and the need for guidelines for its further exploitation, and on the other hand the many changes which are taking place, not least with respect to electronic publishing and the rise of the Internet. My thanks as ever go to all those people, too numerous to mention here, who have provided advice, guidance and information. The views expressed, however, are solely mine, and comments, observations and suggestions will always be welcome – they can be forwarded care of the publishers.

As a final word on the nature of the subject, it is worth pointing out that there is a considerable affinity between any given piece of grey literature and that most distinguished of canines, the greyhound. The latter is a dog which everyone identifies by a number of well known features (it is tall, slender, has a great turn of speed, and very keen eyesight), but which no one really expects to have a coat which is invariably grey. So it is with grey literature – easy to recognize, but quite hard to define.

Redditch
Autumn 1997

Contents

CHAPTER ONE

The Nature and Development of Grey Literature

Introduction

Whenever concern is expressed at the continually increasing quantity of grey literature, and at the difficulty it presents to librarians, information workers and documentation experts on the one hand, and to readers and users on the other, the reply which is often given is that such publications are not intended to form part of the permanent literature, and consequently any problems that may arise will be purely temporary. Unfortunately, this sanguine view is belied by the overwhelming evidence of the durability of grey literature and by its frequent citation in library catalogues, bibliographies and reading lists of all kinds. Those who prepare and issue grey literature do so because such documents offer a number of advantages over other means of dissemination, including greater speed, greater flexibility and the opportunity to go into considerable detail if necessary. Another important reason why grey literature has attained significance as a separate medium of communication is because of an initial need for security or confidentiality classifications which prevent documents from being published in the conventional manner. Over the years, grey literature has come to constitute a section of publications ranking in importance with journals, books, serials and specifications; this importance is now achieving the recognition it deserves and grey literature is no longer the Cinderella of information sources. Any subjective feelings that grey literature is being used and quoted more and more are borne out by the many independent references to it, and by the emergence of databases devoted specifically to standardizing its identification and improving its accessibility.

Difficulties arise because whereas most of the different categories of conventional literature are subject to well-established systems of bibliographic control, grey literature still poses awkward questions in terms of identification and availability. Quite often the term 'half published' is used to describe such literature. Precisely because grey literature is so amorphous and intended for a wide variety of purposes, it is not obliged to conform to the standards of presentation imposed by the editors and publishers of conventional publications, nor to the rigours of a refereeing system.

A tremendous number of items appearing in the grey literature are initially prepared with a known and limited readership in mind, and often they carry distribution lists as evidence of this. Often, too, copies are numbered individually so that each can be accounted for. So long as items of grey literature remain within their original restricted environment, difficulties of identification and accessibility remain relatively few and quite manageable. Complications begin to occur when the details of such publications reach persons other than those originally envisaged.

This state of affairs can come about through a deliberate policy of announcing to anyone interested the news that publications in the grey literature are now available generally, or it may happen incidentally because documents have been quoted openly and perhaps unthinkingly along with other more easily traceable references. Either way, it is at this stage that grey literature undergoes a transition from the category 'restricted and/or temporary' to the category 'open and permanent', whatever the intentions of the originators may have been. At this juncture, questions of priority can arise, and the issue of whether or not an item of grey literature has a rightful place in the primary sources of information on a subject can become a matter for dispute when the race is on to be first into print. Nevertheless, once examples of grey literature pass into the conventional literature, they tend to remain there and are quite frequently quoted and requoted in the original grey literature form long after they have been transformed into other, more familiar types of publication.

Immediately grey literature documents do become referenced in the open literature, interested would-be readers need to be able to ask for them in the correct manner – not always an easy task when a document may be characterized by a range of identifying symbols such as accession numbers or series numbers or, on the other hand, simply not given any unique identifiers at all. Equally, a library or other organization asked by a reader to supply a document must have ready access to a location. Since, in practice, grey literature is notified to national copyright and bibliographic agencies in an extremely haphazard manner, if at all, the tasks of compiling location indexes and establishing comprehensive collections have to rely heavily on cooperative arrangements.

Because grey literature presents problems of identification and acquisition and also because it has developed into a key constituent in the literature of practically all branches of knowledge, strenuous but varyingly effective attempts have been made to bring the situation under control, and an indication of the degree to which success has been achieved insofar as the reader is able to tap the rich information sources available by identifying and specifying what he or she wishes to see, and the libraries and other agencies for their part are able to deliver, is the purpose of the present work.

Definitions

The first question to address is what exactly does the grey literature embrace in terms of documents? In an earlier book in the present series, Chillag (1985) gives Wood's widely quoted definition of grey literature as 'literature which is not readily available through normal book-selling channels, and therefore difficult to identify and obtain'. (This definition first appeared in a survey of the British Library Lending Division's role in handling grey literature (Wood 1982)).

Examples of grey literature include reports, technical notes and specifications, conference proceedings and preprints, supplementary publications and data compilations, trade literature, and so on. It has to be recognized, however, that not all material in all these categories is grey in the sense defined above: conference proceedings may ultimately be published as books or journals, whilst many official publications eventually become commercially available and easy to acquire through normal book-selling channels. Uncertain availability is only one characteristic of grey literature – others include poor bibliographic information and control, non-professional layout and format, and low print runs.

Van der Heij (1985) has pointed out that some synonyms for the word grey used in the professional press are 'non-conventional', 'informal', 'informally published',' fugitive' and even 'invisible'. He also reminds us that documents may be unconventional in many ways, and many conventionally published documents show greyish aspects.

Some writers have ventured to include patents and standards among the constituents of grey literature, but most people would agree that although such items may not be readily available through normal book-selling channels (and indeed are not expected to be), they are in fact published through well-recognized agencies which observe long established conventions of strict bibliographical control and identification systems. Patents and standards are sometimes mentioned in publications which announce details of new grey literature documents, but for the purposes of this book they are treated as being outside the accepted definitions, notwithstanding the fact that many British Standards were issued in grey covers! The characteristics of patents and standards are described in many guides to information sources, as for example in the work edited by Mildren and Hicks (1996).

The term grey literature has become widely recognized throughout Europe, as evidenced by the German 'graue Literatur', the Italian 'Letteratura grigia' and the French 'litterature grise'. Indeed, in France, Bernier (1996) has noted that grey literature 'seldom bears any resemblance to the concepts of literature as such'. In the United States the expression seems to have taken longer to become accepted, and Bichteler (1991) notes that 'gray literature is a term which originated with British librarians; it has had to supplant the longer established American term reports literature'. There is also the little matter of the spelling difference.

The problem of trying to define items of grey literature is one which is well known to the life scientist – it is rather like trying to define the species. Everyone can recognize a piece of grey literature when they see it, but it is not easy to write an explanation which covers all the exceptions. In the past few years, several everyday desk dictionaries have started to include an entry for grey literature, as, for example, *The Chambers dictionary* (1993), which states 'material published non-commercially, e.g. government reports'. Another way of arriving at a definition is to examine how a major journal with the avowed intent of publicizing grey literature tackles the problem. The journal in question is the British Library Document Supply Centre's *British reports translations and theses (BRTT)*, which in addition to announcing material in traditional areas familiar to users of the reports literature such as biological and medical sciences, chemistry, and the various branches of engineering, is increasingly looking at the humanities, psychology and the social sciences. *BRTT* regularly reports details of market surveys and local development plans and studies, and it is also devoting more and more attention to topical issues of major concern such as the environment and economic changes inevitable with the development of the single European market. All the material listed by *BRTT* is held at the Document Supply Centre, and around 23 000 titles are added each year.

In Germany, the Staatsbibliothek Berlin is also showing a concern for local publications and now classifies 70 per cent of its material on community development and city planning as grey literature, for which it is preparing a special database (Lux, 1995; Rose, 1997). As the list of categories embraced by the term grey literature gets longer, so the general understanding of the concept seems to grow less. At least scientists and engineers and research workers of all types would show some vestige of recognition when asked their views on the nature and usefulness of the reports literature; now, when confronted with the term grey literature, their most frequent reaction is one of puzzlement. Even in the library profession itself, uncertainty about what is meant by grey literature still persists, and only gradually is the term being accepted as a description of what used to be known as reports literature, unconventional literature and literature not available through the book trade. Only time will tell how far the boundaries of grey literature are likely to expand. Certainly, at least four categories of publications spring to mind which possibly did not feature in the idea of grey literature when first conceived, yet most definitely meet the requirement of not being available through normal book-selling channels.

The first category comprises publications issued by pressure groups and similar bodies with a special point to make. Often such organizations need to publish quickly, their funds are limited, and there is no scope for the niceties of sale or return and trade discounts. Every penny of the cover price is needed by the pressure group, and in consequence sales are achieved by direct mail or through specialist outlets. Many pressure groups promulgate their message through a publication or series of publications, make their point (or not, as the case may be), and then cease activity. Other groups grow in stature, sometimes because they receive support, sometimes because they encounter resistance, and the publications which they issue become sufficiently well known and in demand to warrant distribution through conventional outlets. Examples of pressure groups' publications are reviewed in the national press almost every week, and instances are mentioned elsewhere in this book.

The second category embraces privately published material, which can extend from the proverbial slim volume of poetry through carefully researched family and local histories to topical stories presented with a particular point of view. Poetry will probably not end up in grey literature indexes or collections, but histories and special pleadings may do so if they attract sufficient notice. An excellent example is the Lonrho book *A Hero from Zero* (reviewed by Rees-Mogg, 1988), which is an account of the 1985 Harrods takeover by Mohamed al-Fayed. More than 40 000 copies were printed and the work was the subject of a question in the House of Commons (23 January 1989); however, as Rees-Mogg pointed out, 'it could not be bought in the bookshops'.

The third category is alternative literature, a difficult-to-define group of publications described by Atton (1996), who in attempting to clear up what is and what is not alternative, poses two questions - firstly, are the topics or perspectives unknown or marginalised in the main stream; and secondly, are the publishers largely absent from library collections. If both answers are yes, then the publications can be regarded as part of the 'parallel culture', a term coined by the former Czech dissident Vaclav Havel, and referring to publications outside the dominant culture and state control. Examples of alternative literature range from fanzines (i.e. magazines by and for amateur enthusiasts, especially science fiction, pop music and football) to directories and handbooks of an unorthodox nature (e.g. Llewellyn's *Teenage Liberation Handbook*).

The fourth category which bears on the definition of grey literature is ephemera, the collective name given to material which carries a verbal message and is produced by printing or illustrative processes, but not in the standard book, periodical or pamphlet format. Makepeace (1985), in his study on ephemera, notes that 'probably the best of the definitions that have been advanced for ephemera are not really definitions as such, but merely alternative names such as 'non-book material', 'grey literature' or 'miscellaneous material''. The main type of potential grey literature that is confused with ephemera is the one which is called 'minor publications' and 'local publications', i.e. books, pamphlets, newspapers, news sheets or other multipage formats with two important characteristics:

(1) they are produced uncommercially, and made available free through casual outlets;
(2) they are produced for limited distribution, e.g. within a society or a local area.

Most items of ephemera are naturally produced for short-term purposes (bus tickets, timetables, leaflets, posters, and so on) and although collectable in themselves form no part of the grey literature database. Leaflets in fact constitute an important form of ephemera, and their treatment in France as a historical resource has been described by Barnoud (1996). Some local material (a parish history, for example) can have a long-term value, and its incorporation in a permanent record is a matter for careful consideration. Problems associated with the assembly and classification of ephemera for inclusion in national collections have been reviewed Drew and Dewe (1992). Formal access to minor publications is available through H.W. Wilson's *Vertical file index*, a monthly subject index to pamphlets and other printed material from a broad range of public and private sources. The *Index* caters especially for the needs of teachers, students and library staff, and is available in print and online.

No consideration of the definition of grey literature is complete without at least a look at the term 'book trade' itself. In Great Britain, the dominant characteristic of the book trade is its diversity – it produces as many titles annually as the United States with a turnover and a domestic market only one fifth of the size. Without the protection of the Net book Agreement (such protection does not exist in the United States), this diversity was always considered likely to disappear. The Net Book Agreement – which guarantees authors and publishers that books will not normally be sold below cover prices – was at first just a trade agreement introduced in 1900, but it received the force of law in 1962 when retail price maintenance was generally abolished in Great Britain. In recent years, pressure began to mount for the abolition of the Agreement, as the concept of books as just another form of merchandise subject to market forces gathered wider acceptance. The book trade continued to resist the abolition of the Agreement, 'but in its heart was fatalistically resigned' (Sutherland, 1988). Indeed, in view of the ruling in December 1988 by the European Commission that the Net Book Agreement was contrary to the competition policy of the Treaty of Rome, and of the subsequent dismissal by the European Court of Justice of the Publishers Association's appeal against the Commission's ruling, it seemed only a matter of time before profound changes would take effect. In the event, the Net Book Agreement was suspended in the Autumn of 1995 pending decisions in the English and European courts, and immediately a number of booksellers announced plans to start discounting certain titles. The general view, however, was that only 10 per cent of books by volume of sales would have their prices reduced. The Agreement was finally ended in 1997, following an order made by the Restrictive Practices Court. Now that abolition has finally taken place, the change is likely to influence the very notion of grey literature, not least because more and more publications may find themselves outside the book trade in its new, more competitive form.

Development of grey literature

The appearance of the idea of grey literature is not new – Schmidmaier (1986) gives a quotation from the 1920s:

'no librarian who takes his job seriously can today deny that careful attention has also to be paid to the 'little literature' (Kleinschrifttum) and the numerous publications not available in normal bookshops, if one hopes to avoid seriously damaging science by neglecting these'.

Schmidmaeir gives a further example from the 1930s:

'For years existing systems have been supplying users with several categories of grey literature; example: bibliographic registering of new publications not available in bookshops within the framework of the Deutsche Nationalbibliographie since 1931'.

Developments with respect to grey literature in Germany have also been described by Hasemann (1986).

In the United Kingdom, the term grey literature began to acquire general currency in the late 1970s; it received an important endorsement when the considerable problems of acquisition, bibliographic control and availability led the Commission of the European Communities to organize, in cooperation with the British Library

Lending Division, a seminar on grey literature in York in December 1978. The main result of the seminar was to pave the way for the setting up of a grey literature database called SIGLE (System for Information on Grey Literature in Europe). The subsequent development of SIGLE is discussed in the next chapter. The seminar also gave rise to the following paragraph in the *Financial times* for 24 January 1979: 'There is a twilight world where information is 'semi-published' and is called by the experts 'grey literature'. Grey literature, according to a recently established definition, is material which is not issued through normal commercial publication channels – which means it is usually very difficult to gain access to it'.

At this point it is worth trying to answer a frequently asked question: what is the difference between reports literature and grey literature? *Table 1.1* offers some suggestions.

Nature of reports literature

The term grey literature is now used, as noted above, to embrace reports, but long before the term grey literature was coined, the expression reports literature had achieved a wide currency and was well understood even if it was regarded as a difficult area to work in. Indeed, because reports have a very special significance in the research and development environment, many workers in science and technology will readily acknowledge that they understand, in a general way at least, what is meant by reports literature, as will of course workers in the library and documentation fields, but when it comes to the more

Table 1.1   Reports literature and grey literature: some key features

Date term first used	Reports literature 1940s	Grey literature 1970s
Features	Accounts of R&D work sponsored by government agencies, with heavy emphasis on defence.	All types of literature not available through the book trade, including reports.
Bibliographic control	Usually two tier using report numbers and accession numbers.	Haphazard with some use of ISBNs.
Format	Paper copies and microforms, with electronic products increasing.	Printed works, typescripts. pamphlets microforms and electronic products.
Development and status	Well-established format used and understood by R&D community.	Enjoying growing publicity, helped by European SIGLE initiative.
Availability	Details contained in comprehensive abstracting journals, e.g. *STAR*.	Improving thanks to efforts by national document supply centres.

specific question – what is meant by a report? – a number of answers seem equally valid. On a very formal level, a report is a document which gives the results of or the progress in a research and/or development investigation. Where appropriate it draws conclusions and makes recommendations, and is initially submitted to the person or body for whom the work was carried out. Commonly, a report bears a number which identifies both the report and the issuing organization.

Reports, in fact, are characteristically the products of organizations, and will vary widely in style and method of publication. As will be seen later, they may range from a few pages of technical notes to multi-volume works describing the development of large projects. Many users would agree that a report is incapable of strict definition, particularly when considered in relation to similar publications, such as, for example, conference papers. A factor which complicates the issue is the existence of publications which have all the attributes of reports (issuing organization, author, date, typescript format, card covers, etc) but which paradoxically bear the clear legend on the front page 'this document is not a report'. Such warnings are customarily ignored by the user, who assumes that if a document *looks* like a report then it *is* a report, and may be quoted as such.

Another way of arriving at a definition is to consider whether a report is a particular kind of communication medium quite distinct from all other kinds. The essential feature of a report, and in particular a scientific or technical report, is that it aims to convey certain specific information to a specific group of readers. A report can be the answer to a question, or a demand from some other person or organization or agency for information. This in itself is nothing special, for questions are being asked and demands being made for information every hour of the working day, and the answers are conveyed by many other written means, in particular by letters, by notes and by memoranda. As a rough indication, reports will be preferred when the information is of a certain length in terms of pages, has a useful life of a least a few months' duration and is addressed to a number of readers. Letters, notes and memoranda can of course be long in themselves, but generally they will be informally structured and as likely as not carry no identifying serial number or other means of bibliographical control.

It is useful at this point to digress and see what the dictionary says about the Latin word *reportare*, from which the English word report is derived. Basically, the Latin word means to bring back, a concept which raises a number of fundamental points of emphasis for any definition of a report. It implies that a person or a corporate body goes out and gets something it is commissioned to get and carries it back to the commissioning agency. In other words, a report is one result of an assignment. In a report the objective is generally more definite and has a more imperative shaping effect than in any other form of exposition. It is just this emphasis in the use of the document that distinguishes a report most strikingly from other types of expository writing; it is prepared for a designated reader or readers who have called for specific information or advice.

How do these definitions relate to the actual output of documents described as scientific, technical, business or economic reports? The contents of such documents will be in accordance with whatever standards are followed by the issuing body and clearly these will exhibit wide variations. Since, in general, reports are not subject to refereeing, it is no bad thing that some degree of uniformity of content if not quality is imposed by calling for certain features to appear in a report as a matter of course. A step in the right direction has been taken by the British Standards Institution (1986) with the publication of its specification for the presentation of research and development reports. The guidance the standard offers is sound and logical, and it is hoped that reports-issuing agencies in the United Kingdom will take heed of its precepts. So far, however, there are very few signs of this happening in practice. A similar publication has been issued by the American National Standards Institute (ANSI, 1987), whilst specialist organizations also provide help, as for example the Defence Research Information Centre's special publication DRIC SPEC 1000 *Format standards for the scientific and technical reports prepared for the United Kingdom Ministry of Defence.*

In practice, in the major reports series discussed later in this work, thousands of reports follow a standard pattern or some modification of it, and are variously described as studies, notes, evaluations, reviews, state-of-the-art surveys, analyses, and the like. On the other hand, many examples can be found where a report is really the proceedings of a symposium, a select annotated bibliography, a standard specification, a handbook, a set of statistical tables, a census return, and so on.

In 1967 an American Task Group on the Role of the Technical Report (Federal Council for Science and Technology 1968) drew up a taxonomy of technical reports which enumerates the general types encountered (and which incidentally has stood the test of time). The types are:

(1) the individual author's preprint, which may end up as a journal article;
(2) the corporate proposal report, aimed at a prospective customer;
(3) the institutional report, the purpose of which is budget justification and image enhancement;
(4) the contract progress report, the most popular species of technical report in circulation, primarily aimed at the sponsor, but also available to an extensive group of interested persons;

(5) the final report on a technical contract effort, generally the most valuable species in the collection, hall marked by considerable editorial effort;

(6) the separate topic technical report, very close to the journal article, and the legitimate target of journal editors;

(7) the book in report form, typically a review or state-of-the-art survey;

(8) the committee report, the report series descriptions of which follow widely varying codes.

This mixture of documents truly recognizable as reports with documents less obviously so, presents few problems to the user in the course of consulting a major reports series file, but readers unfamiliar with the wide coverage of reports literature could be excused for looking elsewhere for information on conferences, standards or bibliographies. This uncertainty is, of course, one of the reasons for the introduction and acceptance of the term grey literature.

An interesting and thoroughly practical definition of a report emerged during a survey of technical reports in Canadian libraries (Mark, 1970), namely that a report is an item issued by an organization usually as an 8.5 by 11 inch paper document or as a microfiche or microform, frequently identified by a report number. Several of the libraries surveyed said that they considered a publication a technical report even if it had no number but looked similar to reports that did have one. Some libraries also indicated that they included in the reports category miscellaneous material too important for storage in a vertical file – greyness indeed.

The uncertainty has not disappeared with the passage of the years and writers still find it necessary to explain why reports are so difficult to identify, as for example the paper by Calhourn (1991).

One result of the imprecision in deciding just what constitutes a report has been that many users of the reports literature regard as reports doctoral dissertations and the preprint series and meetings papers of major American scientific, technical and engineering societies. For the same reason, translations are often looked upon as reports, especially when individual papers are allotted report-like numbers. Dissertations, preprints and translations do, however, have sufficient distinguishing characteristics to set them apart from reports, even though they still belong to the greater family of grey literature, and their special features are discussed separately in Chapter 5.

If the contents of reports can show such variation, then so can their purpose. Thus, as was noted in the taxonomy above, a report document may be a once-for-all account of a specific investigation, a progress report, a summary of a project condensing a whole series of progress reports, or an annual report in the traditional manner. A report's purpose directly affects its availability, and the question of issuing agencies and organizations and their concern over matters affecting security and confidentiality is dealt with in the next chapter.

History of reports literature

Whereas the origins of the grey literature concept can be traced back to the 1970s, the history of reports literature goes back to the beginning of the century, and coincides almost exactly with the development of aeronautics and the aircraft industry, and, although nowadays reports are issued on a tremendously wide range of subjects, a significant proportion is still concerned with the area now known as aerospace. The series of reports usually accorded the distinction of being first in the United Kingdom is the *R&M (Reports and Memoranda) series* of the Advisory Committee for Aeronautics, subsequently the Aeronautical Research Council, which began appearing in 1909. In the United States, the aircraft industry has been represented continuously by the National Advisory Committee for Aeronautics (NACA), now known as the National Aeronautics and Space Administration (NASA), which issued its first report on *The behavior of aeroplanes in gusts* in 1915. Further information on these important reports issuing agencies is given in Chapter 6.

Some authorities, however, consider that these aeronautical publications were antedated by documents which were reports in all but name, notably the *Professional papers* of the United States Geological Survey, which began appearing in 1902, and the *Technological papers* of the National Bureau of Standards, which were issued from 1910 onwards as separate items with individual paging and identified by the letter T. These papers contained the results of investigations into materials and methods of testing, and were eventually incorporated (in 1928) into the Bureau of Standards *Journal of research*. The National Bureau of Standards has been renamed the National Institute of Standards and Technology (NIST), a body with a set of initials confusingly like those National Technical Information Service (NTIS), of which more later. An early example of an individual report is the document presented by the Wreck Commissioner, Lord Mersey, to a final sitting of the court of enquiry into the loss of the *Titanic*, held on 30 July 1912. The vessel had gone down on 14 April 1912 and within a matter of weeks the investigators had compiled an account of the sinking.

The development of the report as a major means of communication, however, dates back only to about 1941, with establishment on 28 June of that year of the United States Office of Scientific Research and Development (OSRD), whose task was to serve as a centre for mobilizing the scientific resources of the nation and applying the results of research to defence. The basis for the expansion was the realisation that the report was the most suitable way of presenting the results of thousands of research projects necessary to promote the war effort. With the cessation of hostilities, the OSRD was disbanded, but since there was no respite in research and development activities, and consequently no abatement in the flow of reports, there arose an urgent need for a central agency to maintain the bibliographical control system which OSRD had adopted for the identification of the projects in its charge. The result was that in certain areas special agencies were created specifically to organize and disseminate information, especially reports information, which the stimulus of war had enlarged from a trickle to a flood.

In the United States, three distinct but interrelated lines of development can be discerned. Firstly, the Publications Board was set up in 1945 (and incidentally gave identity to series of reports still being used today, namely the PB series). Later, the Board was absorbed by the Office of Technical Services (OTS), the agency which also originally had the responsibility in the United States for making available the series of reports prepared by the British and American teams of experts that visited Germany and Japan at the end of and in the period immediately following World War II (see below). Subsequently, in 1964, OTS became the Clearinghouse for Federal Scientific and technical Information (CFSTI); and then in September 1970, the establishment of the National Technical Information Service (NTIS) was announced, to which the CFSTI was transferred and merged.

Secondly, it proved necessary to set up organizations to process reports which were classified in the military sense, and naturally not available simply for the asking. Two early agencies of this type were the Central Air Documents Office (CADO), which in 1948 evolved out of the Air Documents Research Center and Air Documents Division; and the Navy Research Section (NRS), which arose from a contract in 1946–47 between the Office of Naval Research and the Library of Congress. The missions of both were consolidated into the Armed Services Technical Information Agency (ASTIA), a body founded in 1951 and which, incidentally, gave birth to another major reports series which still flourishes today, the AD series (from *A*stia Document).

Astia was under the operational control of the US Air Force until 1963, when its name was changed to the Defense Documentation Center (DDC) and its administration transferred to the Department of Defense (DOD). A further name change occurred when the DDC became the Defense Technical Information Center (DTIC), discussed in Chapter 12.

Thirdly, in the lines of development, World War II saw the start and rapid expansion of the nuclear energy industry, events which in turn generated a whole new world of reports literature. In 1942, a central indexing service was provided for the nuclear activities known under the code name 'Manhattan District' and conducted by the Metallurgical Laboratory of the University of Chicago. Later, the United States Atomic Energy Commission (USAEC) established a Technical Information Service (TIS), afterwards called the Office of Technical Information (OTI), Division of Technical Extension (DTE), and eventually named the Technical Information Center (TIC). Subsequent developments involving the Department of Energy are discussed in Chapter 11.

In the United Kingdom, the take-off point in the growth of reports literature was marked by the appearance in the late 1940s of the CIOS reports (see below) and their successors. The reports are of historical interest only today and are fully described by Poole (1973) in his account of the Technical Information and Documents Unit (TIDU). The reports were written by investigators who had been selected to examine particular fields of German and Japanese industry. The early investigations were mainly concentrated on seeking intelligence vital for the full prosecution of the war and were carried out by teams of United States and British military personnel working under the auspices of the Combined Intelligence Objectives Sub-Committee (hence CIOS). When CIOS came to an end in 1945, its functions were taken over by the British Intelligence Objectives Sub-Committee (BIOS) for the United Kingdom, and by the Field Information Agency, Technical (FIAT) and its parent body, the Joint Intelligence Objectives Agency (JIOA) in the United States. With this division of activities and the end of hostilities, the emphasis of the investigations shifted from a purely service angle to one of civilian industrial interest. All the teams submitted reports on their findings, and in many cases also arranged for the transfer of the original documents on which they were based. The investigations in Germany came to an end in 1947, and the printing and publishing of the reports was completed in 1949.

TIDU was part of the British Board of Trade from 1946 until 1951, when it was absorbed by the Department of Scientific and Industrial Research (DSIR). DSIR itself disappeared with the creation of the Ministry of Technology (MinTech) and the Department of Education and Science (DES) in 1964. In fact, the responsibility for the acquisition and announcement of reports of interest to industry (apart from those concerned with nuclear energy) has over the years rested with a succession of government departments: notably with the Ministry of Supply, transformed in 1959 into the Ministry of Aviation; with the Ministry of Technology, as already noted; and with the Department of Trade and Industry (DTI), established in 1970 and responsible for the Technology Reports Centre until the Centre's demise in 1981. The outcome of this event is examined in more detail in Chapter 12.

The history of the official responsibility for British nuclear energy reports is equally involved. The Ministry of Supply and the Lord President of the Council were responsible until the creation, in 1954, of the United Kingdom Atomic Energy Authority (UKAEA). Since then the Authority has undergone considerable change and reorganization, managing throughout to produce a steady stream of reports; details are given in Chapter 11. Those documents which are made available to the general public are often announced through the medium of The Stationery Office (TSO), formerly Her Majesty's Stationery Office (HMSO), and so represent an early example of reports being subject to conventional bibliographical control in that they were announced in a government publications catalogue.

In recent years, the output of scientific and technical reports in particular, but also of reports of all kinds, has grown apace, the bulk of which are still American in origin, but with increasing contributions from elsewhere, notably Britain, France and Germany.

Considerable changes have taken place and indeed are still taking place, especially in methods of announcement and dissemination. In addition, special collections continue to be established and the pattern of usage becomes more diffuse. Any account of the history of reports literature is therefore bound to be incomplete because events are still unfolding. In fact, the continuing development of reports literature is nowhere more clearly demonstrated than in the frequent changes made to the organizations which are officially responsible for reports. This is particularly so with regard to government departments and agencies, and many such developments are noted elsewhere in this book.

Bibliography

No introductory chapter on the origins of grey literature, especially reports literature, is complete without a mention of some of the basic books and papers on the subject. Many are now considerably dated, but all have a value as general background information and give an insight into conditions as they evolved. The main titles concerned are:

Butcher, D. (1991) *Official publications in Britain* 2nd edn. London: Library Association Publishing

Fry, B.M. (1953) *Library organization and management of technical reports literature.* Washington, DC: Catholic University of America Press

Hernon, P. and McClure, C.R. (1988) *Public access to government information* 2nd edn. Norwood, N.J.: Ablex Publishing Corporation (see especially Chapter 10 – *Technical report literature*, by G.R. Purcell)

Nurcombe, V.J. (1997) (ed.) *Information sources in official publications*. East Grinstead: Bowker Saur. (A guide to official publishing in most countries of the world)

Weil, B.H. (1954) (ed.) *The technical report: its preparation, processing and use in industry and government.* New York: Reinhold

Woolston, J.E. (1953) American technical reports; their importance and how to obtain them. *Journal of documentation*, **9**, (4) 211–219

References

American National Standards Institute (1974) *Guidelines for the formation and production of scientific and technical reports.* ANSI Z 39.18:1974. For details of the 1987 revision, see Drener, R.A.V. Scientific and technical reports: the American National Standard. *Bulletin of the American Society for Information Science,* **13** February/March (35) 1987

Atton, C. (1996) *Alternative literature: a practical guide for librarians.* Aldershot: Gower

Barnoud, M. (1996) Ephemeral literature and history sources: the leaflets in the Bibliotheque Nationale de France. *Bulletin des bibliotheques de France,* **41**, (3) 26–29

Bernier, G. (1996) Literature and parapublications. *Documentation et bibliotheques,* **42**, (1) 37

Bichteler, J. (1991) Geologists and gray literature: access, use and problems. *Science and technology libraries,* **11**, (3), Spring, 39–50

British Standards Institution (1986) *Specification for the presentation of research and development reports.* BS 4811: 1972 (1986). London: BSI

Calhoum, E. (1991) Technical reports demystified. *Reference librarian,* **32**, 163–175

Chillag, J.P. (1994) Grey literature. In *Information sources in physics,* (ed.) D.F. Shaw, Chapter 19. London: Bowker-Saur

Drew, P.R. and Dewe, M. (1992) Special collection management: the place of printed ephemera. *Library management,* **13**, (6) 8–14

Federal Council for Science and Technology (1968) *The role of the technical report in scientific and technical communication.* Washington: COSATI: PB-180944.

Haseman, C. (1986) Grey literature: examples of cooperation. *Zeitschrift für Bibliothekswesen und Bibliographie,* **33**, (6) 417–427

Lux, C. (1995) Local government and consultants: a grey literature database on local reports in the Senatsbiblothek Berlin. In Farace, D.J. (ed.) (1996) *Proceedings of the Second International Conference on Grey Literature, Washington 1995.* Amsterdam: TransAtlantic Publishing

Makepeace, C.E. (1985) *Ephemera – a book on its collection, conservation and use.* London: Gower

Mark, R. (1970) Technical reports in Canadian university libraries. *ASLA bulletin,* **34**, (2) 47–50

Mildren, K. and Hicks, P. eds. (1996) *Information sources in engineering, 3rd ed.* London: Bowker Saur.

Poole, L. R. (1965) The technical Information and Documents Unit. In Staveley, R. and Piggott, M. (eds.) *Government information and the research worker.* London: Library Association

Rees-Mogg, W. (1986) A secret best seller. *The Independent,* 8 November

Rose, G. (1997) Government information in Berlin. *Library Association Record,* **99**, (6) 323–324

Schmidmaier, D. (1986) Ask no questions and you'll be told no lies; or how we can remove people's fear of grey literature. *Libri,* **36**, (2) 98–112

Sutherland, J. (1988) Are books different? *Times literary supplement,* 2 December

Van der Heij, D.G. (1985) Synopsis publishing for improving the accessibility of grey scholarly information. *Journal of information science,* **11**, 95–107

Wood, D.N. (1982) Grey literature – the role of the British Library Lending Division. *Aslib proceedings,* **34**, (11/12) 459–465

CHAPTER TWO

Grey Literature Collections and Methods of Acquisition

Quantities

The problem of defining grey literature spills over into any attempt to try and estimate the number of items already in existence and the quantities added annually. Even the figures quoted by the various abstracting services directly concerned with a category of grey literature relatively easy to count, namely reports, do not always make a meaningful aggregate because many items are announced and abstracted in more than one publication. A very careful and time-consuming analysis would be required to arrive at a total based on such sources. Any estimates which are quoted in respect of the total output in any one year tend to be expressed in very round figures indeed. In the past, some attempts have been made to assemble figures relating to reports. For example, the famous *Weinberg report* (President's Science Advisory Committee 1963) gave a figure of 100 000 government reports issued each year in the United States, and this total has often been quoted since. Confirmation of this vast quantity can be obtained from the study on the scientific and technical information explosion made by Emrich (1970), which gave the following data:

> Semi-formal media (i.e. preprints, technical reports and memoranda) announced in 1967: *US government research and development reports* 45 000 items; *Scientific and technical aerospace reports* 31 000 items.

A further figure is obtained from *Defense research and development of the 1960s* (Defense Documentation Center, 1970), a cumulation of citations and indexes to over 450 000 reports accessioned to the DDC and announced in the decade 1960–69 in the *Technical abstracts bulletin (TAB)* and in *US government research and development reports (USGDR)*.

The introduction to the second edition of *Dictionary of report series codes* (Godfrey and Redman, 1973), complied by the Special Libraries Association, notes that at the time of the appearance of the first edition (1962), estimates of the number of reports issued annually varied from 50 000 to 150 000. Since then, guesses have ranged much higher, but it is very difficult to ascertain a reliable figure. Moreover, if grey literature as a whole rather than just reports literature is considered, the problem, as Chillag (1982) has observed, becomes one of 'how long is a piece of string?' Chillag suggested that 'depending on definition parameters, in a world-wide all-subject context, it is quite reasonable to talk of an output of 100 000 to 200 000 such items per year'. Certainly, the number of bodies issuing reports has grown: the third edition of the *Report series codes dictionary* (Aronson, 1986) records over 20 000 reports series codes used by nearly 10 000 corporate authors.

The difficulties involved in obtaining a realistic figure for reports issued from a simple count of reports numbers allocated have been demonstrated by a variety of studies - see, for example, the comments by Hardwick (1987), who cites earlier studies and also gives five examples of her own. Users' own experience is also a good guide to quantities. For instance, Paul Kennedy, interviewed about his book *Preparing for the Twenty-First Century*, (Kennedy 1993) is quoted as saying, with reference to his use of a personal research team: 'It is a reflection of the vastness of the literature. I simply could not control it all myself. The single issue of global warming, for example, has spun off 25 000 reports'.

Availability and security restrictions

In general, grey literature suffers from availability problems because it is often difficult to find out just what can be acquired, but for many categories there are rarely any security restrictions. This is not the case, however, with the traditional reports literature. There can hardly be a more frustrating experience, in view of the wealth of material so adequately documented, than having a request for a report, sometimes after considerable time and trouble have been taken to identify it correctly, returned to the applicant marked with the words 'not available'. Sometimes a reason may be added, such as 'limited distribution' or 'authorized applicants only'; sometimes no explanation at all is provided. In either case, the would-be reader has to face the fact that he or she has reached the end of the line as far as his or her request is concerned. On the face of things, the availability of reports should present few problems. Special organizations have been set up in Great Britain, in the United States, in Germany and elsewhere, whose prime purpose is to deal with requests for reports literature. These organizations normally give a speedy service extremely efficiently, so that when failures to supply do occur, the applicant is genuinely baffled.

The reasons for non-availability are legion, and part of the problem may simply be incorrect or incomplete identification – many a reject becomes easy to obtain once a check is made to rectify a vital omission such as an accession number. These factors of course also influence success with grey literature as a whole. More serious difficulties arise when non-availability is due to security classifications imposed by government departments. The classifications are sometimes called protective markings, consciously or unconsciously referring to the natural phenomenon of the similarity of some animals' markings to those of their natural surroundings, so helping to prevent their being seen by their enemies. Typical categories affecting reports are 'secret', 'restricted', 'confidential', 'not for publication', and 'proprietary'. The actual degree of protection which attaches to each security marking will vary from organization to organization. In the context of reports literature, security generally means safeguarding and protecting classified documents against unlawful dissemination, duplication or observation because of their importance to national defence or security or to the continuing competitive advantage of commercial organizations. The term 'classified' refers to the degree of secrecy which prevents disclosure to unauthorized persons. In practice, each document is security classified individually and is subject to regular review to determine whether a change in status, or even declassification, is justified.

Usually, the most stringent classification systems apply to issuing departments, but equally organizations and firms entitled to receive classified documents, in order for example to fulfil official contract obligations, are required to observe the same strict security measures on their own premises, and regular checks are made to see that they do so. As a rule, before classified information can be released, the intended recipient must be security cleared. Furthermore, the facility where the classified material will be received, stored and used must also be cleared, and there must be a clearly demonstrated 'need to know' for the information in connection with any task being performed. Advice on the regulations governing security classified documents is best sought by direct contact with the appropriate issuing organizations.

In the United Kingdom, the security of government reports is ultimately covered by the Official Secrets acts 1911, 1920 and 1939. The origins of this legislation have been vividly described by Cook (1985), who echoes a widely held view that Section II of the Act has aroused the most controversy. In fact, the 1972 report of the Franks Committee suggested that Section II should be repealed and replaced by an Official Information Act. According to the Franks report, there are 2000 differently worded charges which can be brought under section II, but the two basic criminal offences under it are the imparting of unauthorized information and its receipt. Secrets are classified under four headings:

(1) top secret;
(2) secret (i.e. information and material whose unauthorized disclosure would cause serious injury to the interests of the nation);
(3) confidential;
(4) restricted.

The 1988 White Paper outlining proposals for change to Section II was claimed by its critics as being more concerned with how to control information more efficiently rather than how to provide the public with more information. Such matters have a direct relevance to grey literature, especially as one of the words used to describe the publications which it embraces is 'fugitive'. The Official Secrets Act 1989 (Chapter 6) received the Royal assent on 11 May 1989 and is described as 'an Act to replace Section II of the Official Secrets Act 1911 by provision protecting more limited classes of official information'.

Apart from formal classification under the Official Secrets Acts, the British government can resort to other measures which make it difficult to get hold of reports, the contents of which have often been widely leaked or publicized. It can, for example, restrict access to a document by depositing a copy in the House of Commons Library (as happened with the report on a salmonella epidemic drawn up jointly by the Ministry of Agriculture, the Department of Health, and the British Egg Industry Council in 1988; and with the report on lead levels in drinking water, drawn up by the Drinking Water Inspectorate and announced in 1992).

The government can also restrict copies of reports to 'essential people' (as happened with the Department of Transport's report on the loss of the *MV Derbyshire* in the Pacific).

In the United States, a somewhat different situation obtains, because after the Watergate affair and the resignation of President Nixon in 1974, several important acts were passed. The 1966 Freedom of Information Act was already in existence, and it had established the principle that people had a right to know. No longer was it up to them to prove why they wanted the information – its was up to the government to prove why it could not be given. The 1974 Privacy Act opened up individuals' files to enable people to inspect and correct them; it also blocked access by others.

The 1976 Freedom of Information Act proved even stronger, because under it 'any person may request information held by any executive department, government or corporation, government controlled corporation or other establishment in the executive branch of government, including the Executive Office of the President or any independent regulatory agency'. The bodies caught by the Act include the Central Intelligence Agency, the Federal Bureau of Investigation, the Treasury, the Food and Drug Administration, the Internal Revenue, and so on. No reason for an inquiry for information need be given unless it directly affects another person, when a 'balance of interest' principle will be taken into account. There are inevitably some exceptions, and if a department claims information is withheld for defence or foreign relations reasons, a court considering the case must be satisfied that such information is properly classified to prevent damage to those interests.

The 1976 Act is believed to have worked something of a revolution – it has made the Federal government move (especially in the areas of consumer issues) and has exposed government wrong-doing. It has certainly encouraged the investigative journalist, who now has greater access to the grey literature. For example, a piece in the British press (Hay, 1988) about biological warfare quoted a US Army technical report *Biological vulnerability assessment: the West Coast and Hawaii*, written by William Rose and Bruce Grim. The correspondent pointed out that the report was classified until he managed to get sections of it released under the Freedom of Information Act.

A further example of the effective use of the legislation is the disclosure of details of an intelligence report issued by the US State Department warning of a planned attack on PanAm Airlines less than three weeks before the Lockerbie air disaster in December 1988. The document was obtained by PanAm's insurers under the Freedom of Information Act some seven years later (Norton-Taylor, 1995). The Act was also instrumental in gaining access to e-mail messages, which some authorities dispute as being documents within the meaning of the legislation. An example of the struggle in this area has been described by Blanton (1995).

The Fund for Open Information and Accountability, New York, has published detailed instructions (termed a 'kit') on how to make a request under the Act, and has provided guidance on monitoring progress, making sure of one's entitlement, checking for completeness, and the lodging of an administrative appeal. In addition, the National Security Archive, which was established in 1986 as a repository for declassified information from the various agencies of the United States government, advises on the most effective ways of filing for documents under the Freedom of Information Act.

In Great Britain, the clamour for a similar Freedom of Information Act continues without respite, and the case for reform is frequently argued by the Campaign for Freedom of Information, which regularly cites instances of documents being withheld in areas as far apart as pharmaceuticals and education. Some redress is offered in the *Code of practice on access to government information* (Office of Public Service, 1996) which includes five commitments to:

(1) supply facts and analysis with major policy decisions;
(2) open up internal guidelines about departments' dealings with the public;
(3) supply reasons with administrative decisions;
(4) provide information under the Citizen' Charter about public services, performance, complaints and redress;
(5) respond to requests for information.

A brochure 'Open Government' explains the *Code* and gives a list of Departmental Openness Contacts. A similar *Code of practice* on openness in the National Health Service came into effect in June 1995.

In fairness, it must be noted that the administrators of security classified collections are constantly striving to eliminate those documents in their charge which no longer merit security protection, largely because the expenditure figures for maintaining inventories of classified information and the associated time-consuming issue and return procedures are very high. Sometimes the documents so released have a very wide general appeal, as for example the data made public by the US National Oceanic and Atmospheric Administration on the topography of the ocean floor (Letts, 1995). The information was originally classified because it was of benefit to submarines operating during the Cold war, but civilian beneficiaries now include weather forecasters, fishermen and mining companies searching for fresh oil fields. Regrettably, classification of one sort or another will always be a characteristic of certain types of grey literature, and all the reader not entitled to security clearance can do is exercise patience, for the information sought may be released one day.

Issuing agencies

The term 'issuing agencies' originally referred to the various bodies initially responsible for the preparation and distribution of reports. Their very multiplicity and the fact that they did not use normal commercial publishing channels to make their output available was one of the impelling reasons for the establishment of secondary centres, specially designed to handle reports received from all quarters. The picture has changed somewhat with the recognition that unpublished or hard-to-obtain material is not by any means confined to reports, which are simply one component, albeit a very large one, in the greater area now known as grey literature.

In the United Kingdom, bodies which typically issue grey literature include:

Associations	Libraries
Churches	Museums

County councils
Educational establishments
Federations
Government departments
Institutes
Institutions
Laboratories

Pressure groups
Private publishers
Research establishments
Societies
Trade unions
Trusts
Universities.

Such bodies have a standing invitation to contribute their publications to the collection which is being built up at the British Library Document Supply Centre, Boston Spa, discussed below. In many cases, however, there is no need to use a secondary distribution centre such as the British Library to obtain items of current British grey literature; it is simpler to apply direct to the original issuing body or disseminating organization. Some bodies make available regular lists of new publications and invite direct applications for copies of recent documents. Others publicize their offerings in catalogues and indexes prepared by a variety of organizations willing to act as selling agents or distribution points.

As to the United States, the most fruitful procedure for a person seeking details about American grey literature, in particular reports, is to try some of the guides and abstracting publications noted below in this chapter and elsewhere in this book.

A body which has had a great influence on the organization and availability of reports literature for many years past and whose principles are still current today is the Committee on Scientific and Technical Information (COSATI). COSATI, whose governance was assigned to the National Science Foundation in 1971, but which is now defunct, was made up of representatives of the Federal agencies responsible for operating scientific and technical information systems. The structure and workings of the committee were made public through COSATI annual reports, and its aim was the orderly and coordinated development of Federal agency information programmes in the public interest.

A major COSATI contribution to the organization of reports literature is the system of subject category lists, a uniform subject arrangement for the announcement and distribution of scientific reports which are being issued or sponsored by Executive Branch agencies, and for the purposes of managing reporting. The system is a schedule consisting of major fields with a further sub-division of the fields into groups. The COSATI *Subject category list* (1964) has enabled abstracts, citations and the like to be gathered into broad fields or groups for display to the user, and also for distribution purposes.

The *Subject category list* has stood the test of time and has been widely adopted and in some cases adapted throughout the world as the basis of arrangement for a number of abstracting and indexing services in grey literature.

National approaches

Since so much national effort has gone into providing and administering grey literature, it is not surprising that national collections and specialized announcement journals have been created to ensure full and economic exploitation. How far they have succeeded or are succeeding is not easy to establish. A study by Klempner (1968) investigated the distribution pattern in the United States for the abstracting and indexing services of the Department of Defense, the National Aeronautics and Space Administration, and Department of Commerce; the results revealed a distribution reaching 27 per cent of the educational and non-profit research centres, 31 per cent of industrial research laboratories, and less than one per cent of manufacturing establishments – a situation regarded as being far from adequate.

Moore and others (1971) studied the difficulties encountered by users of the National Technical Information Service's indexing and abstracting facilities, and found that the main problems were:

(1) coverage too broad for a single index;
(2) duplication of material indexed or distributed elsewhere;
(3) multiple number series;
(4) inconsistency of bibliographic entries;
(5) coverage unpredictable.

As if this were not enough, a comprehensive study conducted by McClure and co-workers (1986) into the effectiveness of a sample of United States academic and public libraries as linking agents, or intermediaries in the provision of NTIS information services and products revealed a number of barriers which were identified as limiting academic and public library staffs and their clientele in obtaining effective access to NTIS and its resources. Some of the important barriers included:

(1) library staff had a low awareness of specific NTIS information services and products;
(2) library staff had limited reference knowledge and skills related to specific NTIS information services and products;
(3) academic and public libraries offered minimal physical access to NTIS materials;
(4) library staff and selected clientele perceived the cost of NTIS information services and products as excessive;
(5) NTIS had unclear marketing objectives and delivery strategies for the provision of its services and products to academic and public libraries.

No doubt similar barriers are perceived by academic and public libraries in the United Kingdom.

A review of the aims of NTIS has come from NTIS itself in the form of a paper by Caponio and Bracken (1987) on the role of the report in technological innovation. In the paper the authors identified a particular type of user of the NTIS services, namely the technologist, or research and development practitioner, whose information needs differ widely from those of the basic or 'frontier' research worker. The authors observe that the technology transfer process always starts with at least a minimal amount of pertinent information. The technologist has to learn that a desirable technology exists, that information on it is available in accessible format, and that there is a procedure for acquiring, evaluating and possibly using it. Such technology transfer has been accomplished in large measure through the medium of the technical report, and of course for over half a century NTIS and its predecessors have served as the primary source for the collection and distribution of government sponsored research and development reports. A more recent inside view of NTIS as a self-supporting agency of the US Department of Commerce is provided by Caponio and MacEoin (1991). All this serves to highlight the validity of the oft-made statement that grey literature is a difficult area, not least because of the perplexing task of identifying who the real users are or should be.

In the United Kingdom, the largest collection of grey literature is to be found at the British Library Document Supply Centre (BLDSC), Boston Spa, and a breakdown of the types of material requested from BLDSC shows the growing importance of grey literature. The figures are serials (English language) 68.24 per cent; serials (foreign language) 3.81 per cent; books (English language) 17.39 per cent; books (foreign language) 0.82 per cent; and grey literature 9.74 per cent. Stocks held are shown in *Table 2.1.* BLDSC houses over 2000 miles of roll microfilm, whilst holdings of reports in microfilm total over 4 100 000 items, with an annual intake of over 140 000 items. The reports collection embraces some 12 000 report series, mainly of NTIS, NASA, AIAA, USDoE, ERIC and INIS items, and is especially strong in aerospace, defence, energy and education.

Table 2.1 BLDSC grey literature stock 1997 (Source: British Library)

Category	Holdings	Annual intake
Reports in microfilm	4 100 000	140 000
Other reports	555 000	30 000
Doctoral theses (US)	462 000	5000
Doctoral theses (UK)	127 840	9840
Conference proceedings	368 590	17 590
Translations	569 645	5645

In 1995, a National Reports Collection was established at Boston Spa and henceforth all British reports literature acquired by the British Library is being added to the collection. The National Reports Collection aims to cover reports, papers and technical notes from private and public sector organizations, charities, action groups and research institutions which are not primarily for commercial sale. Potential users are being urged to make use of the collection to keep up-to-date on the impact and implications of developments in their respective fields. Issuing organizations are also being encouraged to deposit their reports with the Collection to ensure reaching a wider audience. The nature of the problems of identification and acquisition the new collection is intended to overcome have been described by Watson (1996).

Many countries throughout the world now have organizations charged with making arrangements for the handling and recording of grey literature. In Europe, the list comprises the national centres contributing to or associated with the SIGLE database, described below. At one time material emanating from Japan could be accessed in the United Kingdom through the Japanese Information Service in Science Technology and Commerce, part of the British Library's Science Reference and Information Service (SRIS). The service was set up in 1985 and offered access to:

(1) online searching, including Japanese databases;
(2) over 3500 Japanese scientific technical and commercial journals;
(3) over 9 million Japanese patents;
(4) market and industry surveys, company business and trade information;
(5) conferences and reports;
(6) translations, journals and indexes.

The service was disbanded in 1993.

Details of some of the services available from SRIS and elsewhere are provided by Clough (1995) in an introductory guide to Japanese business information. In France, the Institut de l'Information Scientifique et Technique has 'La cellule JAPON de l'INIST' which similarly gives access to and help with Japanese sources. In Japan itself, the Japan Information Center of Science and Technology (JICST) is the main organization concerned with grey literature and its file JICST-Eplus is available on STN International. Citations and abstracts are in English and the material indexed includes public reports, especially of research conducted by the Japanese national government, local governments, public service organizations and public councils. JICST-Eplus replaces the JICST-E database, and the coverage is 1985 to date with close on 3 million records. The file JGRIP contains descriptions of research projects planned or ongoing in Japanese public research institutions.

The Japanese approach to the collection and dissemination of grey literature has been outlined by Ogawa (1995), who describes in detail the work of JICST and the distribution of Japanese scientific and technical information overseas. Japanese activities in the grey literature field have also been described in two reports available through the National Technical Information Service in the United States, namely PB-88 227 780 (*Current status of science and technology; grey literature in Japan)* and PB-90 185 305 (*Grey literature acquisition in Japan).* United States sources of Japanese information include input from the Japan Ministry of International Trade and Industry (MITI), and three reference works available through NTIS: *Directory of Japanese technical resources in the US* (PB-91 100 958), *Directory of Japanese databases* (PB-90 163 080) and *Directory of Japanese technical reports* (PB-91 100 986). A list of Japanese databases overseas is available from the Tokyo-based Database Promotion Center (DPC).

Grey literature originating in France is coordinated by the already noted body INIST, which is an integrated scientific and technical information centre created by the Centre National de la Recherche Scientifique to replace the Centre de Documentation Scientifique et Technique (CDST) and the Centre de Documentation des Sciences Humaines (CDSH). INIST produces the file PASCAL, a multidisciplinary database corresponding to the publications *Bulletin signaletique* (1973–1983) and *Bibliographie internationale* (1984 onwards). Sources include journals, conference proceedings, theses, reports, books and patents. The file contains well over 10 million references. In 1994, the French Research and Higher Education Ministry set up the GRISELI Program within its scientific and technical libraries directorate (Comberousse, 1995). GRISELI is intended to collect, index and disseminate grey literature produced in France via a network of organizations acting as GRISELI centres and each is charged with collecting and indexing grey literature in a given field.

In Germany, a major centre of grey literature activity is based in Hanover, where the Technische Informationsbibliothek (TIB), founded in 1959, works in conjunction with the Universitatsbibliothek Hannover (UB), established in 1831. The resulting UB/TIB places special emphasis on the acquisition of unpublished German research reports, foreign reports and conference proceedings, and publications in East European and East Asian languages and in other languages less familiar in Germany. The collections at Hanover are described in a series of *Jahresberichte* (annual reports), and rapid access to the material collected is provided by a group of services under the general heading TIBQUICK.

Russian grey literature (Shraiberg, 1995) is to be covered by a new database called *Rusgrey*, prepared by the Russian National Public Library for Science and Technology, and available through Cambridge

Scientific Abstracts. Records go back to 1989, and the database carries around 4800 abstracts per year on the pure and applied sciences, technology, economics, ecology, environmental sciences, agriculture and medicine.

In Australia, matters relating to scientific papers and technical reports are overseen by the Commonwealth Scientific and Industrial Research Organization (CSIRO).

In Canada, the responsible body is the National Research Council Canada (NRC-CNRC), established in 1916, which operates through the Canada Institute for Scientific and Technical Information (CISTI) in providing access to a collection which includes 54 000 journals and millions of conference papers, monographs, technical reports and translations. Facilities include an online catalogue and an electronic ordering and delivery system.

Governments as publishers

As might be expected, reports and other series of publications issued as the result of work commissioned or supported by governments and governments bodies are to a limited extent announced in the regular journals which give news of government publications in general. This is true in both the United Kingdom and the United States, but in each case only a minimum of detail is provided, generally with no abstracts or expanded indexes. Her Majesty's Stationery Office (HMSO), the British government publisher, was sold to a consortium led by Electra Fleming in 1996, and is now known simply as The Stationery Office (TSO). In the past, HMSO issued daily lists and a monthly catalogue with annual cumulations, which in particular have announced a number of reports series, including those from the United Kingdom Atomic Energy Authority. Sectional lists were also issued, arranged according to the government departments sponsoring the publications concerned. Most of these services are likely to be continued by TSO. The initials HMSO will not disappear completely from the information scene since part of the old organization remains as a residual body with *inter alia* responsibility for Crown Copyright and the registration of Statutory Instruments. An example of the continuing role is to be seen in the proposal to make the papers held in the *Sir Winston Churchill Archive Trust* available on CD-ROMs. HMSO is to charge a licensing fee for the Crown Papers in which it has copyright.

In the United States, the *Monthly catalog of US government publications* is the most comprehensive general index of federal government publications, and it announces numerous report series with an indication of availability if not obtainable direct from the Government Printing Office. The story of government publishing is a subject in itself – on the development of HMSO and USGPO see, for example, the paper by Rogers (1987). The threats to the dissemination of official information posed by the privatisation of HMSO have been reviewed by Picton (1996), who highlights possible restrictions to public access to information gathered at the citizen's expense.

The general point needs to be made that as long as government and official publications are available through a government publisher (e.g. USGPO) or a successor (e.g. TSO), and hence through the book trade, they cannot truly be regarded as part of the grey literature. Indeed, the TSO has its own bookshops in London, Manchester, Edinburgh, Belfast, Cardiff, Birmingham and Bristol, plus a network of agents, stockists and distributors.

Over the years, a practice has been growing in Great Britain for individual government departments and agencies to bypass HMSO and instead issue publications direct, frequently accompanied by announcement and review in the general press. The result has been that in the recent past over half of all British official publications have not been issued under the imprint of HMSO, amounting to many thousands of documents each year from more than 500 different organizations – a very significant amount of grey literature. One of the means of keeping track of material not handled by the government publisher has been Chadwyck-Healey's *Catalogue of British official publications not published through HMSO*, annual cumulations of which go back to 1980. A welcome amalgamation of the indexes to HMSO and non-HMSO publications was the *Catalogue of United Kingdom official publications (UKOP)*, published originally by HMSO and Chadwyck-Healey, and now by The Stationery Office in association with Chadwyck-Healey. *UKOP* contains over 334 000 records and is updated quarterly, drawing on three sources:

(1) The Stationery Office publications;
(2) Non-TSO publications;
(3) Agency publications sold by the TSO, including those of the United Nations.

UKOP is thus a single catalogue of all UK official publications, regardless of their origins, plus a source for information on many international official documents. The question of non-HMSO British official publications, especially with regard to their accessibility as a research source, has been examined by Marshallsay (1986).

Acquisition of grey literature

Grey literature can be obtained on a routine basis by a variety of methods, but two of the most commonly used are:

(1) exchange agreements with other organizations, if permitted;
(2) purchases by subscription or on a single item order basis.

Various facilities are available – for example deposit accounts and prepaid coupons; monthly standing orders for certain categories of documents such as report series; annual subscriptions to the outputs of various agencies, either in whole or in part by specified categories; and special arrangements with certain library supply companies. Major institutions able to afford and justify subscriptions to entire series of publications clearly enjoy the advantages of an automatic delivery service and also completeness of coverage. Organizations and individuals wishing to acquire or purchase single copies of documents, especially low value items in terms of purchase price, may encounter some difficulties, and it is vital to comply with the 'how to order' instructions which many issuing bodies have found it necessary to draw up. It is rare, for example, for the normal book trade outlets to be interested in trying to obtain reports; it is far quicker and more effective to apply direct to the originators, and every effort should be made to observe the various procedures specifically designed to speed the handling of requests, such as payment with order.

In the case of reports, care should be exercised when ordering by accession number only, because some items may be more readily available in another form. Thus a request for an item identified by a NASA accession number may turn out to be a patent specification more easily obtained from the regular source for patents.

The availability of grey literature is always going to be patchy simply because one of the definitions of grey literature is 'documents not available through the book trade'. How the book trade itself is defined in these circumstances is never made clear, except that it generally includes publishers, booksellers and subscription agents. Many items of grey literature can indeed be obtained through the book trade, and enterprising booksellers often go to great lengths to fulfil orders. Some firms even include the category 'grey literature' in their catalogues of new and forthcoming publications. This is particularly true of companies specializing in library supply services. On the other hand, there are publishers who issue titles with a limited or local interest and insist on direct orders, perhaps to avoid having to offer a trade discount to an intermediary. The criteria by which grey literature is selected and acquired will depend on a collecting organization's policies or an individual's specific needs. A major constraint will be the size of the purchasing budget, for some items have a very high cover price, as for example the Business Intelligence report *Pay performance and career development* (1995) which cost £445.00.

Another constraint will be the amount of storage space, for items of grey literature can be bulky and difficult to shelve. An alternative to the purchase of grey literature items is to borrow them, and the following remarks apply in the main to procedures obtaining in Great Britain and Europe. At the risk of labouring the point, the first essential to observe is that any request should specify the correct bibliographical details: in the case of reports, the relevant accession number; in the case of other documents, at least the ISBN where one has been allocated. Many a request has been slow in producing results because although the reader has taken great pains to indicate the originating body, the author, the title, the date and other normally vital pieces of information, the required accession number or order reference has been omitted. Many items of grey literature seem to have a surfeit of identifiers, and by selecting just a few such as those mentioned above, the reader expects his request to be met, especially when compared with the relatively concise descriptions needed to obtain books and journal articles. Librarians and information workers are justly renowned for their detective-like skills in tracing incomplete references, but eventually even their patience tends to wear thin.

The British Library Document Supply Centre has made great strides in streamlining its lending services for grey literature, partly by making its collection as comprehensive as possible, and partly by rationalizing its request processing routines. In the case of reports, instances can occur when specific items are unexpectedly not held, and there may be delays of up to several weeks whilst attempts are made to obtain them. The British Library has also been obliged to point out on occasion that even though reports originate with or are disseminated by US government agencies, they are not always available unless listed in one of the major reports announcement journals and/or released by NTIS or the originating body. Reports so excluded are best sought by recourse to the specialist organizations and libraries in the subject areas covering the topics in question. NTIS has its own special arrangement when it comes to selling its own publications; the Service used appointed agents for countries around the world, and details of the system are described in Chapter 12.

Large organizations handling hundreds of thousands of documents each year aim to provide satisfactory services to their users, but unfortunately they sometimes become victims of their own size and so do not always succeed

Databases for grey literature

Until the arrival of SIGLE (System for Information on Grey Literature in Europe), databases covering grey literature were generally developed to cover specific subject areas such as science and technology, energy and aerospace, with a particular emphasis on the reports literature. SIGLE is different in that it treats a very wide body of information by document format rather than by subject content. Parallels do exist, notably in the treatment of such specialized documents as patents (for example the Derwent range of indexes), and in standard specifications (as, for instance, the Deutsche Informationszentrum für Technische Regeln DTIR Datenbank). Other databases concentrate on specific aspects of grey literature, including translations, conference papers and theses, all of which are mentioned in Chapter 5.

SIGLE (the name has been registered as a trade mark) is produced by the European Association for Grey Literature Exploitation (EAGLE), the secretariat of which is based in The Hague in the Netherlands. The member countries of EAGLE are represented by the organizations shown in *Table 2.2*:

File data for SIGLE extends back to 1980 and the number of searchable records is now 547 169 (1996), an increase of 11 per cent over 1995, with an addition of approximately 46 000 new items per year. The update frequency is monthly. The types of document recorded are reports (28 per cent), dissertations (35 per cent), conferences (5 per cent) and academic or progress reports (4 per cent). The remaining 28 per cent consists of miscellaneous documents of which a high percentage is reports. The subject coverage of the file data breaks down into social sciences and humanities (31 per cent), technology (33 per cent), natural sciences (24 per cent) and biology and medicine

Table 2.2   National SIGLE centres

Country	SIGLE centre
Belgium	Universite Catholique de Louvain-la-Nuève (UCL) Bibliotheque GSH
Czech Republic	Czech Academy of Sciences, Main Library; Central Technical Library Prague
France	CNRS Institut de l'Information Scientifique et Technique (INIST)
Germany	FIZ Karlsruhe and the Universitatsbibliothek und TIB Hannover
Hungary	Technical University Budapest, Central Library
Italy	Biblioteca Centrale Consiglio Nazoinale delle Richerche (CNR)
Latvia	Latvia Academic Library, Riga
Luxembourg	Bibliotheque Nationale, Luxembourg
Netherlands	Koninklijke Bibliotheek, Den Haag; Jupiter, Den Haag; Sociaal-Wetenschappelijk Informatie- en Documentatiecentrum van de Koninklijke Nederlandse Akademie van Wetenschappen (SWIDOC)
Spain	Centro de Informacion y Documentacion Cientifica (CINDOC)
United Kingdom	British Library Document Supply Centre
	Membership applications have been accepted from:
Portugal	CDCT/Junta Nacional de Inv. Cientifica e Techologica
Slovakia	Slovak Centre of Scientific and Technical Information

(12 per cent). From the outset the biggest contributors to SIGLE have been the United Kingdom, Germany and France, with current inputs at respectively 55 per cent, 26 per cent and 11 per cent. Significant inputs are also received from the Netherlands (4.5 per cent) and Italy (3.0 per cent), whilst those from Spain are increasingly rapidly from a low base.

All documents listed in SIGLE are available at least through the national centre which provided their input, and country codes form part of the record number for each entry: IT =Italy, BE =Belgium, and so on. Documents with a country code XE = are available from the Office of Official Publications of the European Communities, although no XE input was in fact received in 1996.

The SIGLE database is available on line via STN International (worldwide) and BLAISELINE and ABES (EC countries). SilverPlatter issues a CD-ROM with half yearly updates. Full details of SIGLE are contained in the *User guide* (1997) and experience of working with SIGLE has been discussed in a number of countries, including Italy (Pagamonci, 1993), Germany (Kluck, 1993) and Hungary (Mesterhazi-Nagy, 1993).

Whereas SIGLE is international in scope, a grey literature database dealing primarily with German language documents is *FTN (Forschungsberichte aus Technik und Naturwissenschaften)*, known until December 1990 as the *FBR (Forschungsberichte)* database. *FTN* is produced by the Fachinformationszentrum (FIZ), Karlsruhe, and includes records on projects sponsored by the German Federal Ministry for Research and Technology (Bundesministerium fur Forschung und Technologie BMFT). The FIZ serves a German industry's technical and business information centre and is sponsored by industrial employers' organizations, professional associations of engineers and industrial R&D centres The annual output stands at about 110 000 documents derived from international scientific and technical publications.

In addition to the *FTN* database, German speaking nations in Europe are well served in respect of certain categories of grey literature by the printed volumes of the *Gesamtverzeichnis der deutschsprachigen Schrifttums ausserhalb des Buchhandels (Bibliography of German-language publications outside the book trade)*, usually known for short as *GVB*, and published by K.G. Saur, Munich. The work covers the period from 1966–80 and publication, which began in 1988, is still continuing. The sources are the national bibliographies of Germany, Switzerland and Austria, and also the material contained in the *Forschungsberichte*. Entries which were previously set out under various cataloguing rules have been rearranged in a consistent pattern according to the *Regeln fur die alphabetische Katalogisierung (RAK)*. The work is in fact a major series of biblographies on a diverse range of subjects, and includes the *Bibliography of medicine (1993)*,

Bibliography of the performing arts and music (1991), Bibliography of psychology (1992), Bibliography of sport and recreation (1991), Bibliography of language and literature (1992), Bibliography of veterinary medicine (1992), Bibliography of fine arts (1992) and *Bibliography of philosophy and religion (1992).*

Universal availability of publications (UAP)

No account of acquisition problems in relation to grey literature can be considered complete without a brief reference to the Universal Availability of Publications (UAP) programme of the International Federation of Library Associations (IFLA). UAP is both an objective and a programme developed by IFLA with the full support of Unesco. The laudable objective is the widest possible availability of published material (defined as recorded information issued for public use) to intending users, whenever and wherever they need it, as an essential element in economic, scientific, technical, social, educational and personal development. To work towards this objective, the programme aims to improve availability at all levels, from the local to the international, and at all stages, from the publication of new material to the retention of last copies, both by positive action and the removal of barriers.

The Universal Availability of Publications programme has been seen as a major element in a wider concept of universal access to information, and it clearly has important implications for grey literature, especially the linking of bibliographical control and availability at the national level. The early activities of the British Library Document Supply Centre within the framework of the programme have been reviewed by Vickers and Wood (1982), and later, broader developments are covered in a special issue of *IATUL quarterly* (UAP, 1991), which examines among other topics document ordering and supply, and the role of booksellers and agents.

References

Aronson, E.J. (1986) *Report series codes dictionary*, 3rd edn. Detroit, MI: Gale Research

Blanton, T. (1995) *White House e-mail: the top secret computer messages the Reagan/Bush White House tried to destroy.* New York: New Press (W.W. Norton)

Caponio, J.F. and Bracken, D.D. (1987) The role of the technical report in technological innovation. 5th International Conference of Scientific Editors, Hamburg. Report PB 87-232 500

Caponio, J.F. and MacEoin, D.A. (1991) The National Technical Information Service: working to strengthen US information sources. *Reference librarian*, **32** 217-277

Chillag, J. (1982) Non-conventional literature in agriculture – an overview. *IAALD quarterly bulletin*, **27** (1) 2-7

Clough, R. (1995) *Japanese business information: an introduction* . London: British Library

Comberousse, M. (1995) Grey literature in France: the GRISELI program. In Farace, D.J. (ed.) (1996) *Proceedings of the Second International Conference on Grey Literature, Washington 1995*. Amsterdam: TransAtlantic Publishing

Committee on Scientific and Technical Information (COSATI) of the Federal Council for Science and Technology (1964) *Subject category and subcategory structure*. AD-612 200

Cook, J. (1985) *The price of freedom*. London: New English Library

Defense Documentation Center (1970) *Defense research and development of the 1960s: cumulated citations and indexes to defense-generated technical information – a user's guide*. Alexandria, VA: DDC

Emrich, B. R. (1970) *Scientific and technical information explosion*. AD-717654

European Association for Grey Literature Exploitation (1997) *SIGLE user guide*. The Hague: EAGLE Technical Committee

Godfrey, L.E. and Redman, H.F. (1973) (eds.) *Dictionary of report series codes*, 2nd edn. New York: Special Libraries Association

Hardick, M. (1987) *A guide to locating technical reports in US government publications collections*. ED-287506 Metrodocs Monographs One

Hay, A. (1988) Cloud on everybody's horizon. *The Guardian*, 13 September 1988

Kennedy, P. (1993) *Preparing for the twenty-first century*. London: Harper-Collins

Klempner, I.M. (1968) *Diffusion of abstracting and indexing services for government-sponsored research*. Metchen, N.J. : Scarecrow Press

Kluck, M. (1993) *The importance of social science literature as part of the grey literature and the representation of social sciences in SIGLE*. In Farace, D.J. (ed.) (1994) *Proceedings of the First International Conference on Grey Literature, Amsterdam 1993*. Amsterdam: TransAtlantic Publishing

Letts, Q. (1995) Sea map reveals ravines that dwarf the Grand Canyon. *The Times (London)*, 25 October 1995

McClure, C.R., Hernon, P. and Purcell, G.R. (1986) *Linking the US National Technical Information Service with academic and public libraries*. Norwood, N.J. : Ablex Publishing Corporation

Marshallsay, D.M. (1990) *Non-HMSO British official publications*. London: Economic and Social Research Council, ESRC/HOO 23/2037

Mesterhazi-Nagy, M.L. (1993) *Concept of joining SIGLE by Hungarian libraries*. In Farace, D.J. (ed.) (1994) *Proceedings of the First International Conference on Grey Literature, Amsterdam 1993*, Amsterdam: TransAtlantic Publishing

Moore, L. *at al.* (1971) *Distinction is all – NTIS from a technical librarian's point of view* ED-058 913

President's Science Advisory Committee (1963) *Science government and information*. Washington: USGPO (the Weinberg report)

Norton-Taylor, R. (1995) US report reveals Lockerbie warning. *The Guardian*, 27 July 1995

Office of Public Service (1996) *Code of practice on access to government information*. London: HMSO

Ogawa, H. (1995) JICST's acquisition of grey literature published in Japan. In Farace, D.J. (ed.) (1996) *Proceedings of the Second International Conference on Grey Literature, Washington 1995*. Amsterdam: TransAtlantic Publishing

Pagamonci, A. (1993) *The Italian grey literature in SIGLE in the scientific-technical fields*. In Farace, D.J. (ed.) (1994) *Proceedings of the First International Conference on Grey Literature, Amsterdam 1993*. Amsterdam: TransAtlantic Publishing

Picton, H. (1996) Access to official publications : threats and responses. *Refer*, **12**, (2) Spring 8–12

Picton, H. (1996) The privatisation of HMSO. *Refer*, **12**, (3) Autumn 2–14

Rogers, C. (1987) HMSO, USGPO and the history of government publishing. *State librarian*, March 5–8

Shraiberg, Y. (1995) New types of and access to grey literature databases generated by the Russian National Public Library for Science and Technology. In Farace, D.J. ed (1996) *Proceedings of the Second International Conference on Grey Literature, Washington 1995.* Amsterdam: TransAtlantic Publishing

Universal Availability of Publications (1991) Special issue devoted to UAP. *IATUL quarterly*, **5**, (4) 219–276

Vickers, S. and Wood, D.N. (1982) Improving the availability of grey literature. *Interlending review*, **10**, (4) 125–130

Watson, D. (1996) New report collection founded. *Library association record*, **98**, (3) 118

CHAPTER THREE

Bibliographical Control, Cataloguing and Indexing

Introduction

Bibliographical control – the identification and description of documents – may be exercised in a variety of ways according to one of several systems, especially in the allocation to publications of uniquely identifying numbers. Grey literature has always been criticized for the complete absence or inconsistent application of any means of bibliographical control. Indeed, it has often been remarked that one of the characteristics of grey literature is that any given document may assume several different guises, to the confusion of both user and supplier.

Bibliographical control

A system for the bibliographical control of published books by means of book numbering has been in operation for many years and provides for the construction of an International Standard Book Number (ISBN) consisting of ten digits made up from the following parts:

(1) group identifier (i.e. national, language, geographical or other convenient group);
(2) publisher identifier;
(3) title identifier;
(4) check digit.

The purpose of an ISBN is to identify one title or edition of that title if there is more than one, or volume of a multi-volume work, from one specific publisher, and it is unique to that title or edition or volume.

In the United Kingdom the administration of ISBNs is carried out by the Standard Book Numbering Agency, which was set up by collaboration between J Whitaker & Sons (publishers of *The Bookseller*), the British National Bibliography, and the Publishers Association. The Agency's duties are to allocate identifiers to publishers and to see that they create ISBNs satisfactorily; to allocate ISBNs in full to those titles which are published by organizations or persons not producing their own numbers; to register and record all ISBNs and to make them available for bibliographical and listing services. The most comprehensive listing of publishers and their ISBNs is to be found in the *Publishers international ISBN directory (1995/96)*, which contains over 340 000 entries from more than 200 countries and is published annually. Details are given of over 267 000 publishers with ISBN prefixes, plus thousands of publishers with no ISBNs assigned.

In recent years, another identifier for goods of all kinds has come into wide use, namely the European Article Number (EAN), and such an identifier can often be found printed on the back cover of a book as a bar code with an eye-readable number attached, with the ISBN number printed as well.

A system for serial publications has also been developed and involves the use of an International Standard Serial Number (ISSN) (Unesco, 1973) which is a seven digit number plus a check digit written in the form XXXX-XXXX. The number is used to identify a serial title and is inseparably associated with that title. Any change of title requires the allocation of a new ISSN.

Grey literature comprises a mixture of documents, some with ISBNs, some with ISSNs, and very many with neither. In the case of reports, the application of ISBNs has not been very common (examples are the various series of reports on atomic energy announced in the catalogues issued by The Stationery Office). In the United States, most technical reports sent to USGPO depository libraries receive due bibliographical control by virtue of their being indexed in the *Monthly catalog*. Many of the technical reports featured in the *Monthly catalog* are also NTIS reports, but most NTIS documents are not in the depository category, and so do not appear in the *Monthly catalog*.

Over the years, many systems of report numbering have been tried and refined, and in consequence the main means of bibliographical control in the reports literature has been the report number, a simple concept which ought to result in report identifiers that are complete, consistent, concise and unique.

Regrettably, because report numbering is a task assumed by those who handle reports as well as those who issue them in the first place, the situation surrounding report numbers has developed to the point where the Special Libraries Association has been obliged to use one word to describe it: *chaotic* (Godfrey and Redman, 1973). The remedy may lie in standardization, and indeed attempts have been made in this direction. However, before examining what is involved in standardization, it is as well to see what happens when an issuing body decides to allocate a number to a report. The body in question sets out with the laudable intention of being as thorough as possible, and so considers that a report should have all or at least some of the following elements:

(1) symbols for the name of the agency;
(2) symbols for the subject matter of the report;
(3) symbols for the form of the report;
(4) symbols for the date;
(5) symbols for the security classification;
(6) symbols to show that additional data has been added by recipient;
(7) symbols which uniquely identify the report;
(8) symbols which include the location within an organization.

The outcome, of course, can be an excessively long number which is inconvenient to quote verbally or in writing and greatly susceptible to transcription errors such as the substitution of wrong letters, the transposition of numerals, and simple omissions or duplications. The length to which a report identifier can run is shown by the following example, chosen quite at random from *British reports translations and theses*, namely 92–09–03A-001 'The large combustion plant directive; an analysis of European environmental policy'. 1991 DSC:6232.005765 (OIES-WP-EV-7).

Another extreme example, this time from the United States, is to be found in the list of further reading attached to the article 'Can nuclear waste be stored safely at Yucca Mountain?', by C. G. Whipple (1996). The entry reads 'Total systems performance assessment – 1995: an evaluation of the potential Yucca Mountain repository'. Civilian Radioactive Waste Management System (Management and Operating Contractor) B00000000–01717–2200–00136, Las Vegas, 1995.

The best agency to assign a report number is the issuing or originating agency. The next best are the contracting or assigning agencies, but the trouble here is that a project may be sponsored by more than one agency, each of which wants to choose its own identifier. Finally, a report number may be assigned by a recipient. In national collections using widely recognized series, this procedure, far from being a problem, is a positive advantage in that often the recipient's number (also called the accession number) is all that needs to be quoted. Consistency demands that the various codes assigned at different stages in a report's progress be correlated, a reasonable enough requirement, except that in terms of file maintenance and cross referencing, the work involved can become excessively time consuming and costly.

The general use of an agreed format for the creation of unique report numbers would enable issuing bodies to allocate numbers to their publications which would be compatible in arrangement with those assigned by others. A standard is available for the numbering of technical reports (American National Standards Institute, 1983) wherein a technical report is defined as a document that gives the results of research or development investigations, or both, or other technical studies. The word technical is used to mean practice, method, procedure, theory, etc., in any science, art, business, trade, profession, sport or the like. The Standard Technical Report Number (STRN) consists of two essential parts:

(1) a report code, to designate the issuing organization or corporate identity and in some cases a series or special series issued co-operatively by two or more organizations;
(2) a sequential group, characters which constitute that portion of the STRN assigned in a sequence by each report issuing entity.

The maximum number of characters permitted in an STRN is 20, but an optional local suffix is also provided for; it is not part of the STRN and may be of any length. Other standards available include ISO-5966–1982 *Presentation of scientific and technical reports* and ISO-10444-(D13) *International standard technical report number (ISRN)*.

Nevertheless, even if satisfactory provision as outlined above is made for series of reports, translations and other documents issued in long runs, the problem of handling individual items originating from a myriad of different sources will still remain. Certainly, the allocation of ISBNs offers one simple solution, but this step requires a proper depository system, and the very nature of grey literature would appear to preclude anything like one hundred per cent compliance with this principle.

The experience of the Fachinformationszentrum (FIZ), Karlsruhe (Keil and Lankenau, 1994) highlights the problems associated with an adequate descriptive system for grey literature, where it is pointed out that because grey literature is not used as frequently as, for example, journal articles, economic considerations influence the depth of treatment given to such documents.

Cataloguing and retrieval

In one sense, the cataloguing of grey literature does not require special treatment; it can be dealt with under the arrangements so carefully made for literature in general, of which it is a small but growing part.

However, once again reports provide the exception, mainly because for many years they have been regarded as outside the mainstream of library material. Where only small collections of reports are concerned, such documents may conveniently be treated as part of a library's ordinary pamphlet holdings, but in cases where many reports are involved, practical problems may arise owing to the wide range of bibliographical information appended to each document. Standard library cataloguing rules may cease to be practicable and an organization must decide whether the time and effort involved are worthwhile and whether it is not more advantageous to adopt one of the sets of instructions drawn up by reports handling agencies.

Some libraries find that their reports collections do not need cataloguing at all, due to the thorough coverage provided by abstracting journals, and as a consequence simply make do with the provision of an elementary report serial number card or record for each item held. The main exception to this procedure will be the treatment of internal reports arising from the work by the staff in the library's parent organization, such as a research institute or laboratory.

Internally produced technical reports have long been a major information resource in the library of a commercial or industrial organization. Such documents are written for internal circulation only and consist typically of engineering, production, sales development, process engineering and research department reports. Usually they include laboratory notebooks, progress reports, interim reports and final drafts. Thus an organization will collect external reports, which can be treated in one of the ways outlined above, and it will also accumulate its own sub-species of grey literature, which in effect will constitute a valuable corporate resource, and which will require some degree of cataloguing or other formal organization.

A standard scheme for the descriptive cataloguing of government scientific and technical reports has been evolved by COSATI, and subsequently revised (Guidelines, 1986). The aim of the scheme is to provide rules for descriptive cataloguing appropriate to the needs of information centres, documentation centres and the reports handling departments of libraries; to provide users with a consistent form of citation and index entries among the various information systems; to enable government agencies to use each other's descriptive cataloguing entries with a minimum of editorial revision; and to provide a guide for other organizations.

According to the COSATI scheme, the essential descriptive cataloguing elements are:

(1) accession number;
(2) corporate author;
(3) title;
(4) descriptive role – subtitle or progress report, for example;
(5) personal author;
(6) date;
(7) pagination;

(8) contract number;
(9) report number;
(10) availability;
(11) supplementary note;
(12) security classification.

Each of the above elements is carefully defined and by far the largest amount of attention is devoted to the corporate author, described as the institutional or corporate body which has prepared and/or is contractually responsible for a report. Organizations most likely to be involved in issuing scientific and technical reports are identified by COSATI as:

Academies
Arsenals
Associations
Business corporations
Centres
Colleges
Companies
Councils
Establishments
Firms
Foundations
Government agencies
Groups

Hospitals
Institutes
Institutions
Laboratories
Museums
Observatories
Proving grounds
Schools
Societies
Stations
Universities and their foreign (i.e. non-US) equivalents.

The list is markedly similar to those drawn up by people attempting to define the scope of grey literature. The corporate author is sometimes referred to as the source or originating agency. On the other hand, the government or other agency which is financially responsible for the report and looks after its distribution is sometimes referred to as the controlling or monitoring agency. The COSATI scheme specifies that only two organizational elements may be chosen from those displayed on the title page and that they should be the largest followed by the smallest, as for example:

Largest element and place name; smallest element:
General Electric Company, Cincinnati, Ohio; Nuclear Materials and Propulsion Operation.

Special rules apply for city and state names, departmental committees, abbreviations and changes of name. Exceptions to the largest/smallest rule occur when one of the subordinate elements:

(1) includes a proper name or a personal name, e.g. Bureau of Mines, Morgantown, West Virginia, Appalachian Experiment Station;
(2) is an independent name, e.g. Atomics International, a division of North American Aviation, Inc.;
(3) is designated as the responsible organizational level by a report series number.

Even with the availability of careful rules for guidance, the identification of the elements to be chosen is not always self-evident, as for example with the report 'Respiratory health effects of passive smoking: lung cancer and other disorders', the title page of which reads:

United States Environmental Protection Agency, Office of Research and Development Washington, Office of Air and Radiation, Washington EPA/600/6–90/006F December 1992.

This emphasis on corporate authorship in reports literature is in marked contrast with traditional cataloguing procedures, where for many years the personal author was supreme, although nowadays the title of a work is normally taken to be the main entry feature for a catalogue entry. The contrast is highlighted by the massive *Corporate author authority list* (Kane, 1987) prepared by NTIS. It is the printed version of the Corporate Author Authority database created and maintained by NTIS, and contains more than 40 000 main entries for US Government sponsored research, development and engineering reports, as well as foreign technical reports and other analyses prepared by a whole range of agencies, their contractors and grantees

With respect to titles (item 3 in the list of elements noted above), a paper on the criteria for United States technical reports (Hoshovsky, 1965) provides a good example of a title which is a little too concise. The title reads ELF INVESTIGATIONS and is not, as might appear, an inquiry into the denizens of Middle-earth, but a document dealing with the subject of extremely low frequencies. Hoshovsky suggests that the essential elements of an informative title are:

(1) the object – what area is studied?
(2) the purpose – what are we looking for?
(3) the nature – report on an experiment, state of the art review, and so on.

Using these guidelines, the rendering ELF INVESTIGATIONS is expanded into the much more helpful 'Worldwide magnetic field measurement of extremely low frequencies in the atmosphere'.

Another example of the expansion of a short title is the British Geological Survey's report 'Midas project' (1995) which has the informative sub-title 'Multidataset analysis for the development of metallogenic/economic models and exploration criteria for gold deposits in Western Europe'.

The announcement journal *British reports translations and theses (BRTT)* uses augmented titles to help readers, as for example 'Code of practice; rights of access . . .' which is expanded as 'disability legislation' (96–11–05R-022).

The original COSATI cataloguing scheme has also been extensively developed and adapted by the Defense Technical Information Center (Martens, 1992) as guidelines for descriptive cataloguing in the data fields for the computer input of technical documents. The DTIC has also published a vocabulary, the *DTIC retrieval and indexing terminology* (1987), the aim of which is not only to be a manual for DTIC to index and retrieve information from its various databases, but also to be an aid to assist DTIC's users in their own information storage and retrieval operations. The title is now the *DTIC thesaurus* (Jacobs, 1990), a basic multidisciplinary subject term vocabulary. The development of subject category schemes for reports literature has been comprehensively reviewed by Zijlstra (1993).

It is possible to argue a case for the use of the Universal Decimal Classification (UDC) as a suitable scheme for the treatment of grey literature. It is, after all, one of the most detailed of the general classification schemes, particularly in the fields of science and technology; it is truly international, with versions in many languages; and its synthetic nature makes it amenable to use in computer systems.

The UDC has undergone a great deal of revision to the schedules, especially in recent years, and is eminently suitable for the classification of individual collections of grey literature, although the adoption of the Scheme by major announcement agencies does not appear to be imminent. (See Chapter 6 for an example of a publication classifying reports by the UDC). The current version of the *Anglo-American cataloguing rules (AACR)* can be used to deal with various kinds of reports under the section on work of corporate authorship, and the rules do indeed recommend that a work that is by its nature necessarily the expression of the corporate thinking or activity of a corporate body be entered under that body. However, it has to be remembered that the *AACR* is a set of rules for the construction of catalogues of general collections and for that reason does not tend to provide for the cataloguing of specific categories of material (although it does now cater for computer files). The 1988 revision continues this policy and specialist material such as the various categories of grey literature will to be catalogued according the general principles found in *AACR*.

A study comparing technical reports cataloguing records in the Library of Congress MARC format using *AACR2* (i.e. the second edition) cataloguing rules and in the DTIC format using COSATI rules has been reported by Burress (1985), who notes that the automated systems and networks available for controlling monographs are not yet available for reports. Three solutions seem possible:

(1) treat reports as if they were monographs;
(2) use the MARC communications format, inserting COSATI-derived information into hospitable fields;
(3) develop special software.

The Deutsche Forschungsgemeinschaft (German Research Association) is supporting a project on the automated cataloguing of scientific and technical material, including grey literature, which is being run by the Technische Informationsbibliothek (TIB) in Hanover, the Gesellschaft für Mathematik- und Datenverarbeitung (GMD), and the Technische Hochschule, Darmstadt. The background to the work has been described by Roth and Schlitt (1991).

The suggested COSATI format for catalogue cards is: *accession number, corporate author* (initial capitals), *title* (all capitals), descriptive note, *personal author* (initial capitals), date, pagination, *contract number, report number*, availability, and supplementary notes. The italicized elements are those which should be put to the left margin for emphasis and ease of marking for filing.

Of the descriptive elements recommended, the following provide useful index entries in an abstracting journal, or useful filing points in a manual catalogue: accession number, corporate author, title, personal author, contract number, and report number. It is fairly easy to decide from the precise definition given in the COSATI scheme what an accession number actually is, but whether or not it is to be used for cataloguing purposes is not straightforward in cases where a document carries more than one accession number. For example, many reports announced by NTIS are identified by both AD and PB report numbers.

A further aid to the cataloguing of reports is the standard practice of including in each document a control sheet or a template which requires basic indexing information to be set out in a uniform manner. Standard instructions are issued for the completion of such control sheets, and if followed the outcome is a comprehensive record which enables any cataloguing department to identify the key pieces of information for its own records. For example, the International Atomic Energy Agency's International Nuclear Information System (INIS) uses an arrangement of worksheets which provides numbered fields for each element of the bibliographical description of a document to be added to the records.

At one time, there was also a practice of inserting in certain categories of reports and technical notes detachable abstract cards, for the convenience of librarians and others who had to maintain card files in a consistent manner. Examples were to be found in the publications of the Royal Aircraft Establishment, the National Engineering Laboratory, and NASA. More and more card catalogues in libraries of all kinds are being converted to online systems, and with respect to grey literature an example is the retrospective conversion of a technical report collection at the Sterling C. Evans Library, Texas, details of which have been reported by Tull (1991), who describes the entire project as it was carried out by contract cataloguers, from the contract specifications through daily procedures to the final cleaning after tape load.

The question of indexing and retrieval systems for grey literature is one which concerns the end user only insofar as it is necessary to understand the protocols required to address and exploit the various databases which cover the material in question. Major databases are considered elsewhere in this book under the appropriate subject fields, but it is appropriate to recall that the main database which caters specifically for grey literature as a format rather than by subject is, as noted in the previous chapter, SIGLE (System for Information on Grey Literature in Europe), available through the hosts BLAISE and STN.

However, given the patchy nature of national inputs to SIGLE, the wideness of its subject coverage, and the general lack of appreciation and understanding of grey literature on the part of the educational, scientific, technical and academic communities, not to mention the library and information profession itself, SIGLE appears to have a long way to go before it achieves the same sort of status and recognition accorded to its longer established companions.

Filing

In determining filing practice the size of the collection of grey literature will be the deciding factor, and if only a few items are held by an organization, they can be filed by whatever method is most convenient and consistent for the main collection of conventionally published material. If, on the other hand, the numbers of items are considerable, there are important advantages to be gained by a system of segregation according to characteristics likely to be familiar and logical to the reader. Thus runs of standard sets of reports, collections of theses, files of transactions, and so on are best filed by whatever accession or other numbers have already been allocated to them. In physical terms, there will of course be a requirement for an allowance of space between individual series to cater for future additions to the collection, and extra complications can arise if some of the items, especially reports, are security classified. In this instance, there is no option but to put such documents under lock and key in specially designated filing cabinets or cupboards or even a properly designed strong room.

In establishing a filing system for grey literature, the advantages of browsability should not be overlooked, even though in many cases the pamphlet-like format means that documents are unlikely to sit comfortably on the open shelves. Where possible, grey literature should be shelved alongside conventional publications. This applies, for instance, to reports running to many pages, particularly those in hard or semi-stiff covers, which resemble ordinary bound books. A case in point in the NASA SP (Special Publications) series where titles like *Joining ceramics and graphite to other materials* (NASA-SP-5052, 84 pp), *Advanced bearing technology* (NASA-SP-38, 511pp) and *Quest for performance; the evolution of nuclear aircraft* (NASA-SP-468, 548pp), substantial volumes all, are most useful to the reader if placed by similar texts in the ordinary literature sequence.

Further practical guidance on the coding and filing of grey literature, notably any official documents from whatever source, is contained in the work edited by Pemberton (1982), which gives detailed accounts of practice in libraries in Australia, Canada, Ireland, the United States and Great Britain.

Obsolescence

The period for which grey literature should be retained as part of a store of information will vary with local conditions and requirements. Any policy will have to take into account the actual use made of the material, the availability of archive copies in collections elsewhere, and the collection's own role (if any) as an archive source itself. The employment of microforms considerably eases the problem of storage and security, as does the digitization of texts to create digital images for online use and possible CD publication.

Grey literature has not been around long enough as a distinct category to have given rise to many studies as to its obsolescence rates, whereas in the technical reports area a number of attempts have been made to determine how long publications can usefully be retained. In particular, high rates of obsolescence were noted in a study by Wilson (1964) at the Atomic Energy Research Establishment, and much more recently by Vickers and Wood (1982) who were moved to say that 'it has been repeatedly mentioned that the demand for grey literature collected by the (British Library) Lending Division is very low'. Whilst a lack of awareness of the value of grey literature is without question one factor, obsolescence or perceived obsolescence could well be another.

Similar trends in a fall off in use are well documented with regard to the conventional literature, especially in respect of the decrease in demand for periodical articles as the information which they contain is gradually absorbed into the general body of textbooks and reference works. A good example of this process is the classic manual *General and industrial management* by Henri Fayol (1841–1925), the principles of which were first disclosed in a French mining journal. The work has since been through many editions to become a standard in its field (Fayol, 1988), and only the keenest student would now go back to the original articles.

Conclusions

Apart from reports, the application of various forms of bibliographical control to grey literature has not until recently attracted a great deal of attention. Knowles (1987) has reported on the bibliographic presentation of grey literature and McClure and others (1986) have observed that libraries will not find it exceptionally difficult to maintain bibliographic control over NTIS material, should it be purchased. For practical reasons, bodies intimately concerned with reports in all their aspects have been obliged to seek their own solutions. To a considerable degree, a measure of success has been achieved, but in some ways the results stress the differences rather than the similarities between grey literature and conventional publications, and so to a degree perpetuate the feeling that the material is 'difficult'.

References

American National Standards Institute (1983) *Standard technical report number (STRN) format and creation*. ANSI Z.39.23 -1983

Burress, E.P. (1985) Technical reports: a comparison study of cataloguing with AACR 2 and COSATI. *Special libraries*, (76) Summer, 187–192

Defense Technical Information Center (1987) *Retrieval and indexing terminology*, 3rd edn. AD-A-176 000

Fayol, H. (1988) *General and industrial management*, revised by Irwin Gray. London: Pitman

Godfrey L.E. and Redman, H.F. (1973) *Dictionary of report series codes*, 2nd edn. New York: Special Libraries Association

COSATI (1986) *Guidelines for descriptive cataloging of reports: a revision of the COSATI standard for descriptive cataloging of government scientific and technical reports*. PB 86–112 349

International ISBN Agency (1995/96) *Publishers international ISBN directory* 22dn edn. Berlin: International ISBN Agency

Jacobs, C.R. (1990) *DTIC thesaurus*. AD-A-226–000 (supersedes AD-A-176 000, above)

Kane, A.V. (1987) *Corporate author authority list*, 2nd edn. Detroit, MI: Gale Research

Keil, U. and Lankener, I. (1993) Description of grey literature: demand for standardization. In Farace, D.J. (ed.) (1994) *Proceedings of the First International Conference on Grey Literature, Amsterdam 1993*. Amsterdam: TransAtlantic Publishing

Knowles, C.M. (1987) *Bibliographical presentation of grey literature*. Report EUR 7139. Luxembourg: Office for official Publications of the European Communities

McClure, C.R. *et al* (1986) *Linking the US National Technical Information Service with academic and public libraries*. Norwood, N. J.: Ablex Publishing Corporation

Martens, G.B. (1992) *DTIC cataloging guidelines*. AD-A- 246 500/3/GAR

Pemberton, J.E. (1982) *Bibliographic control of official publications*. Oxford: Pergamon Press

Roth, G. and Schlitt, G. (1991) Automated cataloguing of specialized literature. *Bibliotheksdienst*, **25**, (8) 1231–1234

Tull, L. (1991) Contract cataloging: retrospective conversion of a technical report collection. *Technical services quarterly*, **9**, (1) 3–18

Unesco (1973) International serials data system. *Unesco bulletin for libraries*, **27**, (2) 117–118.

Vickers, S. and Wood, D.N. (1982) Improving the availability of grey literature. *Interlending review*, **10**, (4) 125–130.

Whipple, C.G. (1996) Can nuclear waste be stored safely at Yucca Mountain? *Scientific American*, **264**, (6) 56–64

Wilson, C.W.J. (1964) Obsolescence of report literature. *Aslib proceedings*, **16**, (6) 200–201.

Zijlstra, B.H. (1993) The history of US scientific and technical information (STI) and its subject categories. In Farace, D. J. ed. (1994) *Proceedings of the First International Conference on Grey Literature, Amsterdam 1993*. Amsterdam: TransAtlantic Publishing

CHAPTER FOUR

Special Means of Distribution – Microforms and the Internet

Introduction

Providers of grey literature have always been ready to embrace whatever technology is appropriate in the dissemination of their material, and in particular have been extensive users of screen based systems. In the beginning, the screens were used to display microform images of the texts of documents, and more recently the screen has been extensively used to access the wealth of data now available on the Internet. Microforms still have an important part to play in the handling of individual items of grey literature, whilst the Internet is going from strength to strength, but with the ultimate format and outcome not yet certain. Screen based databases have also become an important part of the information landscape, and their significance is discussed elsewhere in this book, leaving the present chapter clear to concentrate on two delivery systems for complete texts, namely the old (microforms) and the new (the Internet).

Microforms

As the volume of grey literature, particularly reports, grew it became apparent that there was a need for a method of producing documents which offered a combination of convenient storage, cheapness of reproduction and ease of transmission. One solution lay in various types of microforms. Initially, reports were issued on microfilms and microcards, and at first 16mm and later 35mm microfilm was used to supply early United States reports to libraries in Europe. In most countries on the receiving end, the microfilm was cut into its separate report lengths and then stored in a small canister marked for retrieval and subsequent reading and printing. Microcards too were commonly used for reports literature, especially the 3 by 5 inch size; however, since they were opaque, microcards presented special technical difficulties with regard to enlarging and duplicating, although they could be read easily enough given a suitable piece of reader equipment.

Microfiche has now ousted microcard and microfilm as the prime microform medium for reports and other separately issued documents converted to microform. Microfiche are sheets of microfilm containing multiple microimages in a grid pattern and a title strip with can be read without magnification. The term 'microfiche' seems to have entered the English language in the 1950s and stems from the French word 'fiche', meaning sheet or card. The plural can be fiche or fiches.

Williams (1970), in reviewing papers on the early development of microfiche in France and Germany prior to World War II, comments that it would be rash to speak of the 'inventor' of the microfiche, since it is one of the oldest of microforms. Microfiche are available in two formats: micro-negatives – that is a clear image on a black background, which prints to give a conventional black text on white paper; and micro-positives – that is a black image on a clear background. Some microfiche combine the two formats - see, for example, some of the individual papers in the *SAE transactions* series – but generally microfiche are made available as micro-negatives. It is worth noting that in 1969 the United States Defense Documentation Center (DDC) (now called the Defense Technology Information Center – DTIC) established an experimental policy allowing DDC users to order microfiche in either positive or negative film, since a considerable number of customers had maintained that positive microfiche were more legible on viewing equipment. After a couple of years, however, it became evident that customer interest in positive microfiche did not justify the expense of the service. DDC also made numerous tests in colour processing, but currently colour microfiche are held to be too expensive for large scale, closely costed systems. The extent to which microforms have been accepted as a means of publication and distribution is indicated by the figures relating to the stock of this medium held at the British Library Document Supply Centre in the year 1997:

Reports in microform	4 100 000 (holdings)
Reports in microform	140 000 (annual intake)
Roll microfilm	over 2000 miles
Microfiche (other than reports)	4 250 000 items.

Such numbers confirm the findings of a study by Salisbury (1965) that microforms, and in particular microfiche, offer the following advantages:

(1) accessibility – searches can be localised;
(2) economy- less storage space is required;

(3) clarity – a superior image quality is available;
(4) speed – production and distribution can be quicker;
(5) durability – microfiche have a long life;
(6) security – less administration is needed in maintaining collections.

The specific use of microfiche for scientific and technical reports has been examined by Williams and Broadhurst (1975), and more recently by Sinkule and Moody (1988), who surveyed the main United States Federal agencies which issue reports in this medium. Rouyer (1990) notes that microfilm is still the technology of the future, since it is reliable, durable and inexpensive, and can compare very favourably with optical disc technology. A major European provider of material in microfiche format is the International Nuclear Information System (INIS), discussed in Chapter 11.

Standardization

The path towards a standard microfiche format has been a tortuous one. When, in the early 1960s, microfiche began to be issued on a large scale, two standards were employed, NASA's 5 by 8 inch fiche and the USAEC's 3 by 5 inch fiche. Then in 1964 the Office of Technical Services, hitherto a user of 35mm microfilm, turned to the 4 by 6 inch fiche (104 by 148.75mm), a format endorsed by COSATI. This particular size was capable of carrying 72 image frames arranged in six rows of 12, with a reduction of 20:1. In practice, however, the first row of the fiche was always reserved for eye-legible identifying details, and only on trailer fiche did six rows actually carry microimages. The other main reports issuing agencies quickly followed suit, and for many years thereafter microfiche were normally of this size. The development influenced other organizations both inside and outside the United States, who likewise adopted the microfiche as a prime dissemination medium for reports. The National Micrographics Association (NMA) in the United States (formerly the National Microfilm Association and now part of the Association for Information and Image Management) has always played an important part in standardization, and the first NMA specification for microfiche, designated M-1, was issued in 1963 and revised in 1967. The 1972 version eliminated the 20:1 reduction, 60 frame format and now provides for only one standard size with one format, namely 105 by 148mm with 98 frames at a 24:1 reduction. This size and format have been adopted worldwide (ANSI/AIMM, 1990).

A series of British Standards is available based on drafts prepared by the Photographic Standards Committee and covers microfiche formats (BS 4187) and microfiche legible headers (BS 6627). In 1972, NTIS indicated that it had adopted the 24 by 98 image format of the NMA standard and added that documents announced in *Government reports announcements (GRA)* would in future be issued in the new format. The USAEC and NASA continued with the 20:1 reduction ratio format for the remainder of that year. NTIS pointed out that the Federal Council for Science and Technology had recommended the adoption of the new format, and gradually the new reduction ratio has come into general use for microfiche publishing. The new format had the effect of reducing the image size from 11.75 by 16.5mm to 10 by 12.5mm, but no great difficulties were reported with users' existing equipment, which proved able to cope and gave rise to no great outcry about a slightly smaller image on the screen. Users with automatic step-and repeat equipment for the reproduction of microfiche did have to make some modifications. Boston Spa (BLDSC) began receiving NMA standard microfiche from the United States in July 1972; at that time the demand for enlargements from the older 20:1 fiche was not expected to decline significantly for some years, and the Library installed an automatic enlarger printer specially built to handle the new NMA format. Today, the copying equipment at Boston Spa includes photocopying machines, microfilm cameras, equipment for processing, enlarging, reading and printing microforms, and a range of specialist copying machines producing microform duplicates.

Economic advantages

The economic advantages of microfiche may be summarized as lower initial costs, cheaper copies, low postal charges, low storage costs and low cost copying. A NATO analysis (Vessey, 1970), taking into account the cost of making a microfiche, the cost of a copy microfiche, the selling price of a copy microfiche, the cost of reproducing a paper copy from a microfiche and the postal charges for despatching up to 25 separate microfiche, found that compared with the cost of a document of 50 pages in hard copy, plus postal charges in the production and distribution of 1000 copies of a 50 page report were, not surprisingly, overwhelmingly in favour of microfiche. Microfiche offer large economies in storage space: a single index drawer in a purpose designed microfiche storage cabinet is capable of holding between 1000 and 1500 fiche, that is over 100 000 pages of hard copy equivalent, assuming some of the frames are not used. The British Library Document Supply Centre claims to have had more extensive experience of handling library materials in microform in a wider variety of formats than any other organization in the United Kingdom, and has a wealth of information on the economics of microforms as a storage medium in libraries.

In theory, even greater economies of storage can be achieved with a format called ultrafiche – that is with images at a reduction rate of more than 98×. A number of systems were tried on a commercial basis, including that of NCR which produced 105 by 148mm fiche with images reduced either to 120× or 150×. At the 120× reduction, the fiche contained 70 columns and 30 rows and provided space for 2100 A4 documents. At the 150× reduction rate, the capacity was of course even greater. Perhaps the ultimate in ultrafiche was the publication by NCR of the *Holy Bible* (1245 pages) on a single fiche measuring 50 by 50mm. In practice, the grey literature handling agencies have chosen to stay with 98×.

Reader resistance

Reader resistance to microfiche is high, despite the development and refinement of reading machines. Two major problems are: the user's inability to move rapidly from one from one section of a report to another, and back again, and place book marks at key points; and the inability to compare two or more microfiche documents simultaneously unless extra viewing machines are situated side by side. Resistance also increases with the number of fiche per document. Whereas a short paper or report which fits comfortably on one fiche is just about acceptable, a long document running to several hundred pages and requiring a number of trailer fiche certainly is not. Nevertheless, the trend towards multiple fiche continues, and a good example is provided by the *Documents on British policy overseas, Korea 1950–51* (Yasamee and Hamilton, 1991), which is supplemented by a set of microfiche of some 1000 pages.

The microfiche is, however, able to offer an advantage of its own in that group viewing is possible. Several people can study the same fiche together using a reader capable of projecting the image onto a large screen.

It is essential that microfiche are of good optical quality, both from the user's point of view for ease of reading, and from the reproduction angle, since a good microfiche copy is vital before a satisfactory print can be taken from it by an automatic enlarger printer. Some defects, such as out-of-focus frames and misaligned pages, are due to poor quality control at the filming source. Others derive from the poor condition of the original document and many reports issuing agencies have been obliged to put out warning notices such as 'This document has been reproduced from the best available copy furnished by the sponsoring agency. Although it is recognized that certain portions are illegible, it is being released in the interest of making available as much information as possible'. Other notices in use include 'Reproduced from the best available copy' and 'Copy available does not permit fully legible reproduction'. Such warnings serve to emphasize greyness in the literal sense, and do little to assuage the outraged feelings of the indignant recipient, who is at a loss to know how, in the age of the total quality concept, such substandard originals can be allowed to get into the documentation system. One solution lies in the stricter enforcement of the document format standards, and in adherence to the recommendations for the preparation of copy for microcopying contained in British Standard BS 5444 (British Standards Institution, 1991). Further consideration of the technical quality of microfiche reports is contained in the studies by Horder (1977) and Fernig (1980). An additional factor which works against user acceptance of microfiche is the necessity of having to use a machine at all. Many users are deterred simply by the prospect of being required to walk to and sit down at a desk in a library or a specially designated office to consult a fiche on a viewer. Again, many people in a wide variety of disciplines and professions do a great deal of their employment-associated reading whilst travelling by train, plane or in the back of a car. In such circumstances, a portable viewer would be of great benefit, but there is little evidence of the acceptance and use of such equipment. Many authorities are firmly of the opinion that the most important factor acting against the acceptance of the medium is the limited availability of viewers, and that if the position could be improved in this respect, other means of encouragement could then be used.

A more optimistic note was struck in a study (Christ, 1972) of the Libraries and Information Systems Center of Bell Telephone Laboratories, which had a Technical Reports Center with a primary responsibility for the acquisition, announcement and distribution of externally generated reports to the company's scientists and engineers. Reports were selected from various abstracting journals and then listed in a semi-monthly internal publication called *Current technical reports (CTR)*. The normal response to requests for reports announced in *CTR* was to supply hard copies on loan, but it was decided to use technical reports as a vehicle for evaluating the microfiche format. The test ran for some nine months, during which time requests were fulfilled by microfiche, and at the end of the period one of the main conclusions was that the reading habits of the test group had not been adversely affected by the use of the microform medium. In fact, the data, comments and observations gathered during the test suggested that, in general, technical staff would accept and use microfiche, albeit grudgingly, provided care was taken to design the distribution system to their needs and preferences.

Readers and reader printers

Since satisfactory equipment is such a key factor, it is worth considering some of the technical aspects of readers and reader-printers. Many different types are currently available and the user has a very wide choice – to mention some manufacturers and not others would clearly be unfair – and a decision is best made as to which is the most suitable after consulting buyers' guides, checking independent assessments, obtaining quotations and visiting trade exhibitions. The list of standards relating to microforms in general and microfiche in particular, both British and foreign, is very extensive as is the literature on the availability and evaluation of equipment and methods . Impartial advice is available in the United Kingdom from Cimtech, the British National Centre for Information Media and Technology, founded in 1967 as the National Reprographic Centre for Documentation. Cimtech specializes in document imaging, micrographics records management, publishing systems and related subjects; a good example of its work is the report by Hendley (1983) on the archival storage of microfilm, magnetic media and optical data discs.

A British Standard (BSI ISO 6198) outlines the performance characteristics expected of readers for transparent microforms. The basic requirements in any microform reader are a sharp image and a sufficiently high light intensity, features which can quickly be assessed by a short test in normal room lighting, but with no bright lights or sunlight shining directly onto the screen. Important features to look for in addition to the purely optical ones are facilities for indexing lines and frame numbers, and the ability to rotate the image to view diagrams and other material printed sideways in the original.

Reader printers are similar in principle to machines designed solely for viewing, but permit enlarged format printing of individual frames, usually as A4 sheets. Since reader printers normally operate at a relatively slow speed (of the order of a half a minute per page), any enlarging is best confined to selected sections of the fiche, and the full-size reproduction of entire documents should be entrusted to specialized agencies or local copying bureaux.

The actual making of microfiche from the original paper copy documents requires skilled operators and special equipment, and again is undoubtedly best left to expert practitioners, although here again guidance in the form of standards is obtainable, as for example British Standard BS ISO 6199 on operating procedures. Similarly, the duplication of microfiche involves the use of purpose-built equipment usually requiring a high rate of throughput to justify the capital cost, and again this process is best left to specialized agencies. Great strides have been made in developing microfilm processors capable of producing films to archival standards, and many companies are now beginning to offer digital and analogue A4/A3 universal reader printers.

An enduring medium

Although by definition the final product is small, in some cases very small, the world of microforms is a large one and the end of developments is by no means in sight .An indication of the durability of the medium is evidenced by the annual cumulation from K.G. Saur called *Guide to microforms in print* which covers domestic and foreign publications issued by commercial and non-commercial sources. The work is complemented by a subject guide to microforms in print, and both guides are supplemented on a six-monthly basis. A very good measure of standardization has been achieved, including terminology (see, for example, the *Micrographics vocabulary* British Standard BS ISO 6196), and this has resulted in a degree of stability which has gradually led to the acceptance of microforms as a versatile medium and no longer just a necessary evil. The combination of diversity and familiarity has meant a greater willingness to use microforms, even though most users still point out that the only really acceptable reading aid is a pair of spectacles. Current developments in microforms can conveniently be followed by consulting the insert published on a regular basis in the *FID news bulletin*, the compiler of which is B.J.S. Williams.

The Internet

If microforms constitute the old screen-based system of dissemination, the new is represented by the Internet. The Internet (from *Inter*national *Net*work) is a network of computers linked by telephone lines along which vast amounts of information can be moved, It is not owned or controlled by a single organization and is open to anyone with a personal computer, a modem and the correct software. The Internet was originally established to enable US government departments, universities and other institutions to exchange information and software. It evolved from a Pentagon funded computer network called the Arpanet (from Advanced Research Project Administration). The basis of the network was a series of super computers at four American universities tied by permanently open phone lines, and since then it has spread in an anarchic fashion to link thousands of computer systems from that of the Library of Congress to countless small businesses, schools and individuals. The father of the Internet is widely regarded as Vincent Cerf, an executive with MCI Communications.

The World Wide Web (WWW) has been described as the friendly face of the Internet in that it allows users to find their way about the network. Its inventor, Tim Berners-Lee, was working at CERN in the European Particle Physics Laboratory in 1989 when he saw the need for a global information exchange that would allow physicists to collaborate on research. He brought together the concept of hypertext, links so called because they are more than just words and letters on a screen, the Internet itself, and the web browser, a piece of software which allows a user to find, view and manage information on the World Wide Web.

Using and exploiting the full resources of the Internet is a subject of growing complexity, and the reader is referred to one of the very many guides now becoming available at an ever increasing rate, such as the *Rough guide to the Internet* (Kennedy, 1995), *The Internet for library and information service professionals* (Dawson, 1995), and *The library and information professional's guide to the Internet* (Tseng and others, 1997).Among the many periodicals devoted to the medium, *Internet reference services quarterly: a journal of innovative information practices, technologies and resources* ((1996) Vol. 1, no. 1. Haworth Press) reflects the current thinking of American information professionals.

Grey literature and the Internet

The Internet, which sprang from the very organizations and institutions which have long been producers of grey literature, appears to have had two main impacts on the world of reports and other non-published documents. Firstly, the Internet has been used as a means of disseminating grey literature, and especially fugitive literature, and secondly, the network has been used enthusiastically by grey literature providers and coordinators to publicize their activities and services.

Since the Internet is still in a transitional stage of development, with opinions still divided on its ultimate role, any observations on the influence of the new medium are necessarily subject to qualification as events continue to unfold.

Examples of grey literature which appear on the Internet range from the legitimate and innovative to the unassignable, illegal and sometimes downright offensive. Taking the legitimate and innovative first, a couple of services will suffice to illustrate current trends. Firstly, in the London borough of Wandsworth, it is now possible for individuals to consult via the Internet local council plans and the associated planning documents, and if appropriate input comments and reactions by e-mail. Local planning documents have for many years been listed in section 05V: Urban, regional and transport planning of *British reports translations and theses (BRTT)*, and a considerable quantity of such items is now on file as part of the grey literature. In this instance, the Internet provides immediate if temporary access and the *BRTT* entries represent a permanent record. Another instance of enterprise on the Internet is the *Roland collection, 1996*, a 2000-page guide to 650 films and videos on art, philosophy, architecture and literature. The collection represents an encyclopaedia selected from more than 8000 documentaries, and includes films from 230 directors. The collection is at *http//www.roland-collection.com*.

The Internet is also noted for the circulation of unattributable material, particularly in the United States, but also in South America, Africa and South East Asia. Thus in 1996 the Internet carried the text of a paper attributed to the Federal Bureau of Investigation on the possible cause of the TWA Flight 800 disaster in which a Boeing 747 crashed into the sea off Long Island in July of that year. The paper was widely quoted in the general press, yet no one could or would vouch for its authenticity or provenance. The press took up the story again in 1997 (for instance in *The Times*, 14 March) with details of a 67-page report on the incident, published in *Paris-Match*, allegedly as a scoop. Reporters pointed out that the document has been 'posted for several weeks on the Internet'.

The Internet has also proved a convenient medium for bodies of outsiders who see any form of government authority as the enemy of freedom to circulate a range of inflammatory texts which could be barred from conventional publishing channels. Finally, the Internet is in some danger of becoming synonymous with pornography, for as the results of a study carried out the Middlesex University and presented to the 1995 meeting of the British Association for the Advancement of Science showed, half of all searches routed via a commonly used search engine were aimed at locating pornographic material.

Apart from what may be termed problems of quality control, the material available via the Internet is also frequently laborious to access with long searches likely to run up considerable phone bills, especially if downloading times is included as well. Despite the resemblance of the Internet to the proverbial curate's egg – that is good in parts – the various agencies associated with grey literature have almost without exception established web sites as comprehensive sources of information about resources and services available. In the United States, the National Technical Information Service (NTIS) has a number of sites, including the NTIS Home Page (*http://www.ntis.gov*) which highlights the general activities of the organization. In the Netherlands, the European Association for Grey Literature Exploitation (EAGLE) has a general inquiry point (*http://www/konbib.nl/infolev/sigle/home.html*) for the System for Information on Grey Literature in Europe (SIGLE), whilst in Great Britain comprehensive information about the British Library Document Supply Centre (BLDSC) is accessible via the Home Page at (http://portico.bl.uk/dsc).

In the specific field of engineering, a good example of an Internet enterprise is the concept of the *Ei Village* established in 1995 by Engineering Information Incorporated (Ei). The *Ei Village* offers a total engineering information package, and includes links to sources of technical reports such as the *NASA technical reports server*. Reviews of the early days of the project have been published by Raeder (1996) and Tenopir (1996). Other Home Pages are mentioned elsewhere in this book, and reference is also made to the growing amount of trade literature appearing on the Internet.

The advent of the Internet has inevitably prompted a great deal of rethinking among workers active in the grey literature field, and many aspects are currently being examined and evaluated. Gelfand (1995) sees the opportunity for the redefinition of the merit of grey literature with enhanced options of combining text with graphics and images, whilst Luzi (1995) considers whether the distribution of World Wide Web servers has stimulated both the production and distribution of grey literature as well as changing its 'traditional' features. Cruz and coworkers (1995) have looked at the problems caused by the fact that grey literature is not covered by legal deposit requirements and therefore apparently by copyright law. Heer and Farace (1995) have examined the research and development needed for an international network for the promotion of grey literature, and have reviewed the products and services of Grey Net, the Grey Literature Network Service established in the Netherlands in 1992 to promote and support the work of authors, researchers and intermediaries active in the field of grey literature. Two key activities of Grey Net have been the ongoing compilation in database format of an *Annotated bibliography on grey literature*, and the preparation of the *International guide to persons and organizations in grey literature*. It should also be noted that one facility the Internet offers is the chance to supplement or even to supersede references to documents cited in the traditional manner. Thus the reference to the report *Making the most of our libraries* ((1997) (British Library Research and Innovation Report 53 BLRIC) is supplemented in a published paper (Bryant, 1997) by the statement 'will be available at http:// www.ukoln.ac.uk/blri053'. If the practice spreads, thought will have to be given to the consistent allocation of web site identifiers.

Conclusions

Despite predictions in some quarters that the Internet will go the same way as the CB radio craze of the 1970s, improvements in software will make usage of the system more efficient and more effective. Current estimates put the number of people connected to the Internet as over 50 million, an endorsement for the term often used to describe the Internet, namely the Superhighway. Certainly, the future developments look likely to cater equally for large organizations anxious to market themselves on the World Wide Web, and entrepreneurs and individuals keen to make contact with persons of like mind and common interests in subjects ranging from genealogy to geranium culture.

References

American National Standards Institute – Association for Information and Image Management (1990) ANSI/AIIM M35–1990. *Micrographic microfiche*

British Standards Institution (1988) *Microfiche* BS 4187:1988

British Standards Institution (1991) *Recommendations for the presentation of headers for microfiche* BS 6627:1985 (1991)

British Standards Institution (1991) *Recommendations for the preparation of copy for microcopying* BS 5444:1971 (1991)

British Standards Institution (1991) *Micrographics: microfilming operating procedures* BS ISO 6199:1991

British Standards Institution (1991) *Micrographics vocabulary* BS ISO 6196 1991–1993.

British Standards Institution (1993) *Micrographics: readers for transparent microforms* BS ISO 6198: 1993

Bryant, P. *et al* (1997) Sour grapes and cherry picking? *Library Association record*, **99**, (7) 378, 380

Christ, C.W. (1972) Microfiche: a study of user attitudes and reading habits. *Journal of the American Society for Information Science*, **23**, (1) 30–35

Cruz, J.M.B. *et al* (1995) Grey literature copyright and new technologies. In Farace, D.J. (ed.) (1996) *Proceedings of the Second International Conference on Grey Literature, Washington 1995*. Amsterdam: TransAtlantic Publishing

Dawson, A. (1995) *The Internet for library and information professionals* . London: Aslib

Fernig, L. (1980) *Using microfiche in documentation work*. Geneva: International Labour Organization

Gelfand, J. (1995) Grey literature in new packages: implications from the transition to electronic publishing. In Farace, D.J. (ed.) (1996) *Proceedings of the Second International Conference on Grey Literature, Washington 1995*. Amsterdam: TransAtlantic Publishing

Heer, I. and Farace, D.J. (1995) Research and development of an international network for promoting grey literature: a case study involving the use of the Internet. In Farace, D.J. (ed.) (1996) *Proceedings of the Second International Conference on Grey Literature, Washington 1995*. Amsterdam: TransAtlantic Publishing

Hendley, A.M. (1983) *The archival storage potential of microfilm, magnetic media and optical data discs: a comparison based on a literature review*. Hatfield: National Reprographic Centre for Documentation

Horder, A. (1977) *Technical quality of microfiche reports: a preliminary study*. Hatfield: National Reprographic Centre for Documentation

Kennedy, A.J. (1995) *The Internet and World Wide Web: the rough guide*. London: Penguin Books

Luzi, D. (1995) Internet as a new distribution channel of scientific grey literature: the case of Italian www servers. In Farace, D.J. (ed.) (1996) *Proceedings of the Second International Conference on Grey Literature, Washington 1995*. Amsterdam: TransAtlantic Publishing

Raeder, A. (1996) Finding engineering information. *Searcher*, February, 1–6

Rouyer, P. (1990) Microfilm – the technology of the future. *Bulletin des bibliotheques de France*, **35**, (2) 116–121

Sailsbury, J.T. (1965) *A study of the application of microfilming to the production distribution use and retrieval of technical reports*. Report AD 615 800

Sinkule, K.A. and Moody, M.K. (1988) Technical reports of the US government. *Microform review*, **17**, (5) 262–269

Tenopir, C. (1996) Moving to the information village. *Library journal*, 1 March, 1–2

Tseng, G. *et al* (1997) *The library and information professional's guide to the Internet*, 2nd edn. London: Library Association Publishing

Vessey, H.R. (1970) *The use of microfiche for scientific and technical reports.* AGARD Advisory report 27, N70–39851

Williams, B.J.S. (1970) *Miniaturized communications: a review of microforms.* London: Library Association

Williams, B.J.S. and Broadhurst, R.N. (1975) *Use of microfiches for scientific and technical reports: consideration for the small user.* Hatfield: National Reprographic Centre for Documentation

Yasamee, H.J. and Hamilton, K.A. (eds.) (1991) *Documents on British policy overseas: Korea 1950–51.* London: HMSO

CHAPTER FIVE

Theses, Translations and Meetings Papers

Introduction

The term grey literature by no means meets with universal acceptance, and in the area of theses, translations and meetings papers, three categories of publication with long and honorable pedigrees, the description sits particularly uneasily. The justification for the inclusion of these categories rests on the following factors:

(1) theses translations and meeting papers are quite frequently announced in journals devoted to grey literature, especially reports literature, and so have come to be identified with the genre;
(2) theses translations and meetings papers have many of the format characteristics of report documents;
(3) theses translations and meetings papers are commonly assigned identifying numbers, either by the issuing establishments or the collecting agencies, and sometimes by both – the numbers bear a strong resemblance to report numbers;
(4) theses translations and meetings papers are regarded not without some justification as difficult documents to identify and obtain, problems frequently mentioned in connection with grey literature in general.

Each of the three categories is considered in more detail below.

Theses

A definition of a thesis formulated by the British Standards Institution (1990) is a 'statement of investigation or research presenting the author's findings, and any conclusions reached, submitted by the author in support of his candidature for a higher degree, professional qualification or other award'.

It is immediately apparent from this definition that a thesis (in the United States the preferred term is 'dissertation') has several points in common with a report. Both documents present details of investigations and/or research; both offer findings and conclusions; both are submitted to an overseeing body (the commissioning agency or the university); and both are regarded as unpublished documents, except, as noted below, in the case of some European universities.

British theses

In the United Kingdom, universities have generally adopted a very conservative attitude to the provision of copies of theses, in marked contrast to the arrangements prevailing in the United States and Europe. There is no central supply channel for British doctoral theses, although the British Library Document Supply Centre has succeeded in building up a comprehensive collection going back to 1970.

Additions are announced in *British reports translations and theses*, and currently BLDSC receives theses from virtually all UK universities. BLDSC also has a collection of theses awarded under the aegis of the Council for National Academic Awards (CNAA). In 1996, BLDSC held 118 000 UK doctoral theses, with an annual intake of 6250 items. The Centre also holds 457 000 US doctoral theses, with an annual intake of 5850 items. Another important repository for theses is the University of London, which receives many items not available from Boston Spa.

An important reference work for the verification of details about British and Irish theses, whether held at Boston Spa or not, is the *Index to theses with abstracts accepted for higher degrees by the universities of Great Britain and Ireland and the Council for National Academic Awards*, compiled and published by Aslib on a regular basis since 1953, with entries going back to 1950. Current volumes of the *Index* no longer refer to the CNAA on the title page, and the work is now prepared in conjunction with Expert Information Limited. The *Index* is also available via the Internet.

From the *Index*, which spans the period 1953 to date, it is possible to ascertain what theses have been submitted and then apply direct to the universities in question for permission to examine them. From volume 35 onwards, the *Index* has been published in an expanded and improved version which includes full texts of abstracts and a greatly enhanced subject index. Entries are arranged according to the following subject classification:

A Arts and humanities	G Medicine
B Social sciences	H Agriculture and forestry

C	Physical sciences	J	Mechanical engineering
D	Chemistry	K	Electrical engineering
E	Earth science	L	Civil and chemical engineering
F	Life sciences		

Around 10 000 theses a year are indexed. For British theses accepted before 1950, checks can be made in the *Retrospective index to theses in Great Britain and Ireland 1716–1950*, a comprehensive five volume work published in 1975. The actual availability of theses will vary from university to university, but many establishments have agreed to follow the procedure suggested by the Standing Conference of National and University Libraries (SCONUL), which states:

(a) that at least one copy of every thesis accepted for a higher degree should be deposited in the university library;
(b) that, subject to the author's consent, all theses should be available for interlibrary loan;
(c) that, subject to the author's consent, all theses should be available for photocopying;
(d) that authors of theses be asked at the time of deposit to give their consent for lending and photocopying in writing, and that this consent should be inserted in the deposit copies of the theses.

Finally, it should be noted that many universities require the reader of a borrowed copy of an unpublished thesis to sign a declaration that no information derived from a study of the text will be published without the consent of the author in writing.

Before leaving the subject of British theses, it is important to note that the Council for National Academic Awards, mentioned above, a body established by Royal Charter in 1964 as a self-governing organization to award degrees to students taking courses approved by it in non-university institutions, was dissolved in March 1993. The reason is that academic responsibility has increasingly devolved to accredited institutions which control the academic standards of their own taught courses and research. All the polytechnics, most of which changed their names to include the title 'university' from October 1992, together with a number of colleges of further education, have now been accredited.

United States theses

In the United States and Canada, and indeed throughout the world, nearly four hundred cooperating institutions ranging from the University of Adelaide in Australia to York University, Canada, submit either the full texts or abstracts of doctoral dissertations to the Ann Arbor based University Microfilms International (UMI), who then prepare a monthly compilation called *Dissertation abstracts international (DAI)*. This publication appears in three sections, namely:

Section A – the humanities and social sciences;
Section B – the sciences and engineering;
Section C – European abstracts (issued quarterly).

Each author-prepared abstract, up to 350 words in length, describes in detail the original research project upon which the dissertation is based. Most of the approximately 35 000 dissertations published by UMI each year are abstracted in *DAI* and may be purchased in microform or as paper copies. As with many other databases, the importance of CD-ROM products is now firmly established, and the *Dissertation abstracts* database is also available on two archival discs covering respectively the periods 1861 to June 1980 and July 1980 to December 1984. A disc from 1985 onwards is updated every six months.

Other printed reference tools prepared by UMI include the 37 volume *Comprehensive dissertation index (CDI)* covering the years 1861–1972, which lists more than 417 000 dissertations accepted in North America under keyword headings with a separate author index. The *CDI ten year cumulation* 1973–1982 cites over 350 000 dissertations; the *CDI five year cumulation* brings together in 22 volumes over 175 000 dissertations; and annual supplements complete the picture.

Masters abstracts international, published quarterly, provides 150 word author-prepared abstracts of masters theses, whilst *Research abstracts*, also published quarterly, includes 300 word abstracts of post doctoral and non-degree published research in special areas such as psychology and education. Finally, *American doctoral dissertations*, issued annually on an academic year basis is arranged by subject categories, institutions and authors, and compiled for the Association of Research Libraries.

UMI's Dissertation Information Service is in fact a vast programme of publishing, bibliographic and copying activities, the foundation of which is a comprehensive database accessible via a search system called Datrix Direct, complemented by *Dissertation abstracts online*, which reproduces in full the texts of abstracts published since July 1980.

Theses in other countries

In order to trace theses in other countries, it is necessary to consult the national lists published in the countries concerned. French theses, for example, have been recorded since 1884 in the *Catalogue des theses de doctorat soutenues devant les universités francaises* (the exact title varies over the years), and also in *Supplement D* of *Bibliographie de la France*. In addition, the database *Téléthèses* under the direction of the Ministere de l'Education Nationale and administered by the Centre National du Catalogue Collectif National, provides details of all doctoral theses submitted in France in all disciplines since 1972 (1983 in the field of medicine), with almost 220 000 references at the end of 1991. An appraisal of *Téléthèses* has been made by Ferrier (1990). German theses may be traced through the *Jahresverzeichnis der deutschen Hochschulschriften* and *Gesamtverzeichnis deutschsprachiger Hochschulschriften (GVH)*. In addition to such comprehensive lists, it is possible to examine the lists issued individually by many continental universities – for example *Dissertationen: Rheinisch-Westfalisch Technische Hochschule, Aachen* and *Thèses: Université de Génève, Faculté des Sciences*. Copies of the documents listed in many such compilations are often offered for public sale by publishers appointed by the universities, so, strictly speaking, taking the material outside the bounds of grey literature.

There is a growing number of critical appraisals of theses as sources of information, and of the databases which have grown up around them. For example, in the United states, Boyer (1972) noted that while doctoral dissertations must embody the results of extended research, be an original contribution to knowledge and include material worthy of publication, they are often overlooked in the preparation of bibliographies and reading lists; Billick (1989) looked at the *DAI* database in relation to the humanities researcher, considering both theoretical concerns and practical applications; and Walcott (1991) reviewed the value of theses as sources of information for workers in the field of geoscientific research. In Russia, the quality of doctoral theses has been investigated by Granovsky *et al* (1992), especially with regard to the originality of the information disclosed, and in Germany studies at the University of Bochum (Doktor, 1992) have revealed that the subjects most in demand by library users were medicine and law, whilst theses on the natural sciences were the least requested.

Many theses eventually give rise to books published in the normal way, but tend not to be able to escape from the dead hand of constraints imposed by their academic background, and so have a reputation for dullness. There are some notable exceptions of course, as for instance Paul Taylor's *Dutch flower painting 1600–1720*, which one reviewer (Ekserdjian, 1995) described as 'the least boring book ever to derive from a doctoral thesis'.

Translations

Translations have always been regarded by some as part of the reports literature, and in recent years by practically everyone as part of the grey literature. In fact, of the thousands of translations made each year, relatively few are of the texts of reports; mostly the originals in question are books, papers, standards and patents. A closer look, however, reveals that apart from certain entire volumes and some cover-to-cover translations of journals, most translations are of individual items such as articles from journals in all disciplines, features in newspapers, or sections from books, and in the translated versions many of these items have several attributes of reports noted earlier, namely pamphlet format, identification numbers, issuing agencies, and also some inconsistency of bibliographic control. Moreover, several of the large agencies which have assumed responsibility for collecting, coordinating and publicizing translations are also the same agencies for dealing with reports, or are organizations which are run on similar lines to reports handling agencies and have adopted some of their methods.

Translations have always been an important part of the reports literature, and are now a major constituent of the grey literature. The reasons why translations, especially technical translations, are in such steady demand are not hard to seek, and may be best summed up by reference to some of the conclusions reached after a classic study conducted by the then National Lending Library and reported by Wood (1976) who found that (a) something like 50 per cent of the world's scientific and technical literature is published in languages other than English, and (b) that information contained in languages other than English is vital for the work of English-speaking scientists around the world. The study went on to recommend that published guides and unpublished indexes to translations should be given more publicity, and indeed a number of organizations have taken up the challenge, as will be seen below.

Before arrangements for translations of individual items (papers, parts of books, standards, patents, ect.) are examined, note must be taken of an area where a degree of bibliographical control is especially good, namely translated versions of entire individual books, and cover-to-cover translations of some journals.

Translated books are listed annually in *Index translationum*, published by Unesco. The *Index* enables the reader to follow from year to year the flow of translations from one country or central region to another, and assists in tracing works by a specific author as they appear in translation. The cumulative version contains over 600 000 references dealing with every subject (literature, social and human sciences, natural and basic sciences, art and history), and provides information on more than 150 000 translated authors

A point worth noting with reference to the various journals which announce new translations is that they are tertiary services in as much as the scientist or research worker who is seeking information will have done his or her searching in the conventional primary journals and possibly also in secondary sources such as abstracting publications. It is unlikely that he or she will be interested in seeing what translations are of use by browsing through the appropriate subject categories in translations announcement journals as well. It is far more likely that a worker will get to know of the foreign language papers and reports of interest to him or her through contacts with colleagues, especially those overseas, and his or her main concern will be to confirm that they are available in translation. For that reason, translation announcement journals have become the special province of the librarian and documentation worker, who have the task of making sure the contents of such publications are disclosed to potential beneficiaries.

The pros and cons of cover-to-cover translations have been rehearsed many times – on the one hand they are considered wasteful because they include articles of doubtful or transient value; on the other hand they are regarded as a means of avoiding omissions and certainly simplify checking when searching an extensive run of issues.

Evidence of the system's durability is the fact that over the years more than 1300 cover-to-cover and selective translation journals have been produced, many from Russian originals, but also from other Slavonic languages, and from Japanese, Chinese and German. The list of journals currently being translated cover-to-cover, as recorded in the journal *World translations index*, now stands at over 300 titles. Full details may be found in the irregular serial *Journals in translation*, the 5th edition of which was published by the British Library Document Supply Centre in association with the International Translations Centre in 1991. At the end of 1994, when the production of a 6th edition was being planned, it was agreed that ITC would continue to produce the bibliography independently. The opportunity was taken to move away from the established format of a hardback edition published every three or four years, and since 1996, *Journals in translation* has been available in printed form and on diskette.

In addition to cover-to-cover translations, *Journals in translation* gives details of journals translated selectively and journals which consist of translations of articles collected from a variety of sources. On a more restricted basis, a list called *Japanese journals in English* was published in 1985 by the British Library Lending Division (as it then was) and the British Science Reference Library; it covers the location of scientific, technical and commercial journals. The Special Libraries Association (SLA) *Guide to scientific and technical journals in translation* (Himmelsbach and Brociner, 1972) is still available from University Microfilms International.

An important factor determining the usefulness of cover-to-cover translations is the time-lag between the publication of the original text and the appearance of the translated version: in some cases, the gap has been so great as to warrant commissioning the translation of a particular item considered especially valuable in advance of the cover-to-cover version.

Many thousands of translations of individual pieces of literature are published each year by government departments and agencies, by industrial firms, by universities, and by commercial translators. In the United States, the major announcement publication and retrieval tool was for several years *Technical translations* (1959–1967), issued by the Clearinghouse for Federal Scientific and Technical Information (CFSTI, now NTIS). A change in policy resulted in translations originating in the US Government sector, such as the TT series sponsored by the National Science Foundation's Special Foreign Currency Science Information program, and the JPRS series, discussed below, being announced in the publication *US Government research and development reports* (USGRDR, now GRA & I). At the same time, the National Translations Center (NTC), a cooperative non-profit enterprise founded in 1963 after having existed for a number of years as a volunteer project of the Science and Technology Division of the Special Libraries Association, became a depository and information service for translations originating in the US non-government sector, and began publishing the *Translations register-index*. The *Register* section announced new accessions in the Center, which was based at the John Crerar Library of the University of Chicago. The *Index* section included details of accessions as well as items listed by NTIS, and items available from commercial translations agencies and many other sources. The *Index* sections of *Translations register-index* accumulated quarterly for all entries to date included in a volume. For NTC accessions, the *Register* section listed translations in subject categories according to the COSATI terminology, and the final units of the identifying reference number corresponded to the COSATI classification.

In 1969, the NTC issued an important guide, *Consolidated index of translations in English (CITE)*, with cumulated translations added to the pool prior to 1966 and consolidated translations announced in *Technical translations: bibliography of translations of Russian scientific and technical literature (1953-1966)*; the *SLA author list of translations (1953)* and *Supplement (1954); Translations monthly (1955–1958)*; and *Bibliography of scientific and industrial research (1946–1953)*. Altogether *CITE* contains details of 142 000 translations. Its companion *CITE II*, covering the years 1967–1984, appeared in 1987. Translations continued to be notified in the *Translations register-index* until 1986. Since 1987 they have been included in the *World translation index*, produced by the International Translations Centre (ITC), discussed below. The National Translations Center was transferred from the John Crerar Library in Chicago to the Library of Congress in 1989.

For users who do not have ready access to such major bibliographical tools, a useful guide to translations of scientific and technical literature has been issued by the Naval Ocean Systems Center in the United States (Wright, 1987); this document describes major providers of existing translations and suggests procedures for having a publication translated by a commercial firm or government agency.

A particularly intensive area of translating activity in the United States, which has been in progress for many years, is that conducted by the Joint Publications Research Service (JPRS), which produces many hundreds of pages of scientific and technical translations each year, a large percentage of which were originally published in the languages of the former Soviet Union. Translations of political and economic material from China, North Korea, Eastern Europe, Africa, Latin America and the Middle East are also included.

JPRS is a part of the US Department of Commerce and was set up to provide translations support to the US Government. Its publications are composed of items selected from foreign language sources by various government agencies and departments, and translated and printed by JPRS. Translations are issued on an *ad hoc* basis, for which many reference aids are available (see for instance, report JPRS 58548) giving details of publications issued in specific subject categories during a given year. Translations also appear as regular series in the form of English translations of publications from various geographical areas – for example *JPRS USSR serial reports* and *JPRS worldwide serial reports.*

JPRS translations are notified in the announcement service *Transdex index* and also in various other notification services, including *Government reports announcements and index*, and *Scientific and technical aerospace reports*, where a typical entry reads N92–27931/4/ GAR (JPRS-USP-92–001), giving details of translations into English from various central Eurasian articles on space. The aims and objectives of the JPRS programme were considered in a paper by Morton (1983), and are broadly the same today.

Great Britain does not have a translations centre so named, but until 1984 Aslib maintained a card index of existing and in-progress translations. Known officially at the *Commonwealth index of unpublished scientific and technical translations*, the service offered a purely location facility with records of close on half a million translations from all languages into English.

The success rate in establishing locations was not particularly high, about one inquiry in ten. A valuable adjunct to the *Index* was an Aslib service providing the names of translators expert in given languages and also possessing a knowledge of specific subject areas. In 1984, the *Index* was merged the database operated by the International Translations Centre, and is available to users through that centre.

The British Library began collecting translations and details of translations in the late 1950s and records of holdings now total many hundreds of thousands, including translations prepared by the US National Technical Information Service, the National Translations Center, and JPRS. The British Library Document Supply Centre actively encourages British government departments, firms, institutions, research associations, universities and other bodies to contribute translations to its collection. Several thousand items per year are received from British sources. BLDSC accepts queries about the availability of translations and if it is unable to locate a source, is prepared to check with other agencies.

At one time, the British Library published the *LLU (Lending Library Unit) bulletin* (irreverently known as the *Lulu bulletin*), later to become the *NLL translations bulletin*, but subsequently, as a result of arrangements between the National Translations Center and the British Library, an exchange of translations and information was established between the two organizations. Translations acquired by Boston Spa from British sources were listed in the *Index* section of NTC's

Translations index-register. Current announcements of translations from British sources are now carried in *British reports translations and theses (BRTT)*, and also recorded in BLDSC's *Translation index*. Another example of a national initiative is the *Canadian index of scientific translations* maintained by the Canadian Institute for Scientific and Technical Information, which gives locations for over half a million translations of foreign language documents.

For many years, the British Library operated a Russian Translating Programme whereby translations of Russian language papers and books (generally not more than two years old unless considered of special significance) were translated into English free of charge, provided the organization submitting the request undertook to edit the manuscript for technical accuracy. Subsequently, the approved translations appeared in the RTS series and were announced publicly, at which stage other interested parties could purchase copies at a very low cost.

In 1973, the British Library initiated a modification to its translation programme by asking inquirers whether an English language summary of a given paper had already been seen. The object was to establish whether translation requests could be satisfied in the first instance by the provision of an English language summary or translated captions to figures and conclusions, plus a photocopy of the original item. The experiment showed that a compromise in the form of an English summary plus short notes was acceptable in many cases, and seemed to indicate that perhaps a great deal of unnecessary translating activity was taking place.

In Western Europe (apart from BLDSC), the largest translation inquiry service is run by the International Translations Centre (ITC), at one time known as the European Translations Centre. ITC is a not-for-profit international network inaugurated in 1961 under the auspices of the Organization for Economic Cooperation and Development (OECD). Its aim is to prevent duplication of translating effort by collecting, processing and disseminating information on existing translations in all fields of science and technology. It provides access to translations through the already noted *Journals in translation* and through *World translations index*, a database and printed periodical representing the combined efforts and input of national centres in Belgium, Germany, France, The Netherlands, Portugal, Spain, Sweden and the United Kingdom. The database is accessible via DIALOG, ESA/IRS and NERAC. In addition, ITC receives United States input from what is known as the CENDI Group (the departments of Commerce, Energy, NASA, Defense and Information), represented by NTIS. *World translations index* is produced by means of the PASCAL system developed by CNRS-INIST, and the references it contains are arranged according to subject headings which are derived from the COSATI scheme. ITC began by concentrating on translations from non-Western languages, but has gradually been increasing its source language coverage and now adds to its files details of translations from one European language to another.

There is growing evidence (Risseeuw, 1992 and 1993) that the work of ITC and the contents of the *World translations index* are having considerable value in overcoming foreign language limitations to international trading, although it is also noted that 'the status of translations in the field of grey literature is not clearly defined'.

Printed cumulations of ITC's records are available as *World transindex (1977–1985)* and *World index of scientific translations (1967–1971 and 1972–1976)*.

On the question of the availability of translations, some commercial organizations engaged in the business of translating specifically forbid reproduction by photocopying, and insist on clients making outright purchases of texts. This is because translating is an expensive and labour-intensive occupation, which is likely to remain so until machine produced translations become available; the problems of producing acceptable machine translations have been surveyed many times, as for example by Schneider (1989). Recent developments in the field can be followed by consulting the series *Translating and the computer*, Volume 18 of which was published by Aslib in 1996 and contains the proceedings of a conference held in London in 1995.

Finally, despite the considerable efforts around the world to keep track of and make accessible the translations that have already been made, inquiries to national centres still result in disappointment. The only recourse then is to commission a translation oneself, either using one of the many agencies offering such services, or seeking advice on suitably qualified translators from specialist bodies such as Aslib. A useful guide to the presentation of translations is to be found in BS 4755 (British Standards Institution, 1989), and of course when completed, a copy of the translation should ideally be added to the general pool of grey literature by formal notification of the relevant details to the appropriate national collecting centre.

Meetings papers

Meetings papers or preprints are terms usually applied to the texts of papers made available in advance of or at meetings and conferences, where they may be presented in person by their authors. The practice is especially common in the United States and many large scientific technical and engineering societies issue preprints in advance of their meetings and annual gatherings, and usually each paper bears an identification in the form of a serial code. After the meetings or conferences have taken place, the papers are critically reviewed and a certain proportion selected for inclusion in the organizing body's permanent records. The remainder are simply listed, and although they are not intended to be a part of the permanent literature, they are nevertheless in the public domain and so are often quoted and requested by persons who are not members of the organizing institution. A further complication is that not all of the papers announced before an event and in consequence assigned identification numbers are actually issued in paper format; some may be presented orally without an accompanying text, and some indeed may not be presented at all.

Preprints and meetings papers undoubtedly qualify for inclusion in the grey literature, but ought not to be confused with unpublished reports, since quite clearly they are (or could be) advance copies of journal articles or transactions papers which eventually end up in the conventionally published literature. However, because such documents carry numbers which resemble in some respects those assigned to reports and because they also have a pamphlet format, many users of the reports literature tend to regard meetings papers and preprints as reports as well. Abstracting and indexing services are well aware of this distinction, and the major providers such as *Engineering index* treat preprints and meetings papers as a part of the normal serials literature. Not all the societies which issue preprints can be covered in this section; instead the characteristics of some of the series most frequently met with will be examined.

In doing so, it needs to be borne in mind that recent advances in electronic publishing, especially via the Internet and the World Wide Web, are modifying the traditional role of preprints as part of the process of scientific communication, and the most recent changes are described by Cruz and others (1995), whilst a specific example of the use of the World Wide Web in the area of high energy physics preprints is cited by Kreitz and others (1995).

American technical societies

The American Society of Mechanical Engineers (ASME) issues a widely used series of preprints, which are announced in the Society's journal *Mechanical engineering* and in its programmes for specific meetings. Preprint code numbers indicate meetings by date and event/subject code, e.g. paper number 88-WA/APM-22 means paper no. 22 presented at the Winter Annual Meeting in 1988 and contributed by the ASME Applied Mechanics Division. A selection of papers is subsequently published in the various journals, such as the *Journal of dynamic systems* which go to make up the *Transactions of the ASME*. The *Annual indexes to ASME papers* list all papers, including those not actually selected for publication in the *Transactions*.

The Society of Automotive Engineers (SAE) also has a very active publishing programme, the thinking behind which has been explained by Straiger (1973). SAE meetings are announced in one of the Society's journals, and also in separate programmes for specific meetings. The identification of individual papers takes the form SAE 920900, where the first two digits represent the year of presentation. The remaining digits pinpoint a particular paper, in this case the first item on the programme of the 1992 Annual Earthmoving Industry Conference. Just under half the papers, after careful review by SAE committees, are selected as having the greatest long-term reference value and are published in the normal way in the *SAE transactions*. The remainder stay firmly in the grey literature category, but complete sets of published and unpublished technical papers are made available on microfiche, and individual titles of either category can be ordered as required. A cumulative subject/author/chronological index dating from 1965 is revised on a regular basis and gives access to many thousands of SAE papers.

Other societies which publish meetings papers on a regular basis are the Society of Manufacturing Engineers (SME) with approximately 800 new papers each year; the American Institute of Aeronautics and Astronautics (AIAA), the papers of which are announced in the Institute's *Journal*, and which for many years complemented the reports output from NASA (see Chapter 6); the American Society of Lubrication Engineers (ASLE), which uses its journal *Lubrication engineering* to announce details of its preprints; and the vast output of the American Society for Testing and Materials (ASTM), information about whose publications can be found in the monthly *ASTM Standardization news*.

Before and after

Information about forthcoming meetings and conferences is available from a very wide range of sources, and listings appear in daily newspapers (for example the *Financial times)* and in many weekly and monthly journals. In the United Kingdom, a major coordinating service is *Forthcoming international scientific and technical conferences*, a long established compilation produced by Aslib. The publication provides details of forthcoming UK and international conferences in all fields of science and technology, with more than 1000 conferences noted each year. Entries give date, title and location, with a contact address for inquiries, plus subject, location and organization indexes.

In the world at large it is becoming more and more convenient to consult online databases for news of forthcoming events, because they are up-to-date and offer a very comprehensive coverage. A good example is *Meetings agenda*, which lists forthcoming meetings, congresses, conferences, exhibitions, colloquia and seminars up to three years in advance, providing world coverage in the technical, scientific and social science sectors. It is compiled by the Centre d'Etudes Nucleaires and is available on Questel.

Once meetings and conferences have taken place, the task becomes one of verifying whether proceedings have been published and generally made available. By far the best source is the British Library Document Supply Centre's *Index of conference proceedings*, published monthly with annual cumulations. A microfiche version is available for the years 1964–88. The *Index* lists some 16000 newly acquired conference proceedings annually. Further compilations are available as *Inside conferences* which provides paper level details from the conference proceedings received, and is published quarterly; and *Boston Spa conferences*, which is also quarterly and contains records of over 350 000 conference proceedings held at BLDSC going back to the year 1787. Comprehensive as it is, the BLDSC's *Index of conference proceedings* simply confirms the fact that a conference took place and to find out what went on at the conference it is necessary to be able to identify the subject content of individual papers. This task can be accomplished by looking at *Inside conferences* and at major reference works such as the *Conference papers index*, issued by Cambridge Scientific Abstracts as a print edition on a bi-monthly basis with 72 000 abstracts per year, and as an Internet edition with monthly updates. The *Conference papers index* covers the life sciences, environmental sciences and marine science.

Another finding aid is the *Index to scientific and technical proceedings (ISTP)*, compiled monthly by the Institute for Scientific Information and providing a complete bibliographical account of each proceedings covered. The work indexes approximately 4100 published proceedings each year – over 160 000 individual papers in all.

Conclusions

Of the three types of grey literature discussed above, namely dissertations, translations and meetings papers, the last-named seem to cause the most problems. Dissertations, whilst admittedly not always readily accessible, are nevertheless very well indexed; and translations have given rise to specialist centres and indexes which attempt to control and coordinate their announcement and availability. Meetings papers, however, owing to the provisional and somewhat transient nature of their contents, can be much more elusive to the extent that their acquisition often greatly taxes the ingenuity of accessions librarians and interlending agencies alike.

References

Billick, D.J. (1989) *The DAI database and the humanities researcher – theoretical concerns and practical applications.* In McCrank, L.J. (ed.) *Databases in the humanities and social sciences 4.* Medford, N.J. Learned Information Inc, 71–77

Boyer, C.V. (1972) *An analysis of the doctoral dissertation as an information source.* ERIC Document ED-065 157

British Standards Institution (1989) *Specification for the presentation of translations.* BS 4755:1971 (1989)

British Standards Institution (1990) *Recommendations for the presentation of theses and dissertations.* BS 4821:1990

Cruz, J.M.B. *et al* (1995) Preprints -communication through electronic nets; an example of bibliographic control. In Farace, D.J. (1996) *Proceedings of the Second International Conference on Grey Literature, Washington 1995.* Amsterdam: TransAtlantic Publishing

Doktor, G. (1992) Dissertations: loans and research relevance in different subjects. *Mitteilungsblatt (Verband der Bibliotheken des Landes Nordrhein-Westfalen)*, **42**, (2) 106–119

Ekserdjian, D. (1995) The Godhead in detail. *The Times*, 6 July 1995

Ferrier, A.M. (1990) Téléthèses. CD-Theses: a data bank and CD- ROM for French doctoral theses. *Documentaliste*, **27** (2) 98–101

Granovsky, Y.V. *et al* (1992) Information based evaluation of the quality of doctoral theses. *Scientometrics*, **23**, (3) 361–376

Himmelsbach, C.J. and Brociner, G.E. (1972) *A guide to scientific and technical journals in translation*, 2nd edn. Washington: Special Libraries Association

Kreitz, P.A. *et al* (1995) The virtual library in action: collaborative international control of high energy physics preprints. In Farace, D.J. (ed.) (1996) *Proceedings of the Second International Conference on Grey Literature, Washington 1995.* Amsterdam: TransAtlantic Publishing

Morton, B. (1983) JPRS translations. JPRS and FBIS translations: polycentralism at the reference desk. *Reference desk review*, **11**, (1) 99–110

Risseeuw, M., Wilde, D.V. and Cooper, N.R, (1992) Global awareness boosts productivity. *13th National online meeting*, (ed.) Williams, M.E., Medford, N.J.: Learned Information Inc., 307–313

Risseeuw, M. (1993) Translations, a darker shade of grey – their value and accessibility. In Farace, D.J. (ed.) (1994) *Proceedings of the First International Conference on Grey Literature, Amsterdam 1993.* Amsterdam: TransAtlantic Publishing

Schneider, T. (1989) *Problems of machine translation and semantic knowledge.* In McCrank, L.J. (ed.) *Databases in the humanities and social sciences 4.* Medford, N.J.: Learned Information Inc., 579–586

Walcott, R. (1991) Overlooked sources of information: theses and dissertations. *Compass*, **68**, (2) 104–106

Wood, D.N. (1976) The foreign language problem facing scientists and technologists in the United Kingdom. *Journal of documentation* , **23**, (2) 117–130

Wright, K. (1987) *Translations of scientific and technical literature: a guide to their location.* NASA report N88–23686/4/GAR

CHAPTER SIX

Aerospace

Introduction

The opinion is often advanced that a progressive industrial society needs a pioneering spearhead technology to stimulate growth, promote innovation and set standards of excellence for a nation's entire range of manufacturing and service industries. In many countries, aerospace fulfils this role, even though dissenting voices are raised from time to time to express concern at the colossal expenditures incurred.

In Europe, the organization responsible for the advancement of space research and technology and the co-ordination of national spaces programmes is the European Space Agency (ESA), formed in 1975 out of two earlier bodies, the European Space Research Organization (ESRO) and the European Launcher Development Organization (ELDO). Part of ESA is the ESA Information Retrieval Service (ESA/IRS), which serves more than 10 000 companies, research centres, government organizations and universities. In the United States, the National Aeronautics and Space Administration (NASA) Aerospace Database represents the largest collection of aeronautical and space science information in the world and contains close on 3 million citations to journal, report and related aerospace literature gathered from around the world. Some of the factors which help to account for such large information handling agencies and such huge amounts of literature, especially grey literature in the aerospace field, include:

(1) the immensity of the projects undertaken in terms of cost (for example in November 2003 NASA's Jet Propulsion Laboratory intends to launch a $200 million solar probe to try and find the answers to some fundamental questions about the sun's corona

and solar winds; another costly venture, this time in Europe, is
the Eurofighter project, a £40 billion collaborative effort involving
Great Britain, Germany, Spain and Italy);

(2) the employment of large numbers of highly qualified, highly skilled
personnel;

(3) the importance of aerospace for reasons of national security;

(4) the need to conduct research and development projects on a broad
front simply to maintain a strategic advantage or a competitive
position;

(5) the ruling, at least in the United States, that contractors working
on government funded programmes are required to report on
progress at regular intervals, subject only to security restrictions.

Security has always been a great feature of aerospace information, and
many publications stay inaccessible to those who fail to demonstrate
a need to know. Even if this classified material is disregarded, there
still remains a large body of aerospace documentation which is actively
promoted, firstly for the benefit of the aerospace industry itself by
making available a common fund of knowledge and experience, and
secondly to try and justify and if possible recoup some of the huge
expenditures involved by furthering and encouraging technology
transfer and product spin-off.

Reports issuing agencies

Many agencies throughout the world issue and coordinate reports and
other documents on various aspects of aerospace, and only representative bodies for some of the major countries actively concerned can
be indicated here. For a comprehensive list of agencies reporting in
any one year, the reader should consult the annual cumulative source
index issue to *Scientific and technical aerospace reports (STAR)*, a
publication discussed below, where entries are arranged alphabetically
by the name of the organization responsible for a document's original
publication.

Aeronautical Research Council (ARC)

In Great Britain, the principal agency with a major output of reports
on matters aeronautical was the Aeronautical Research Council (ARC),
a body which under one name or another dated back to 1909 when
the Advisory Committee for Aeronautics (with functions virtually identical with those of the eventual ARC) was appointed to advise the
Prime Minister of the day. In 1920, the body became the Aeronautical
Research Committee, and in 1945 the ARC, until its disbandment in
1980. The Council's function was purely advisory – that is it had no

executive power over the conduct of research and possessed no funds of its own. Moreover, the ARC was concerned purely with research as distinct from development.

ARC published through Her Majesty's Stationery Office two series of reports embodying the results of research. The manuscripts were initially submitted to the ARC by government research establishments, firms and universities for discussion and consideration as to their suitability for publication. The two series were the *Reports and memoranda series (R&M)*, that is, reports having permanent value and printed by a photosetting process; and the *Current papers series (CP)*, of ephemeral interest but occasionally used to secure the speedy publication of data or information of immediate importance, and duplicated by reprographic means. HMSO published a *Sectional List* (no. 8) devoted solely to ARC documents and between them the two series constituted a major communications channel for the results of aeronautical research in the United Kingdom. ARC's system of reports identification could sometimes cause problems, because although ARC documents originated in different organizations, those organizations were not named in the HMSO catalogues or in the *Sectional Lists*. Thus the reader had no means of ascertaining the original number assigned by the issuing agency. It is true that this information could be gleaned from other sources, but the treatment varied according to which announcement journal was consulted. For example, ARC CP 1173 was listed in the HMSO *Sectional List* simply as 'A parallel motion creep extensometer' by J. N. Webb; in *R&D abstracts* the corporate author was revealed as the Aeronautical Research Council, Teddington, with an ARC internal reference number (ARC 32319); whereas in *STAR* the corporate author was stated as the Royal Aircraft Establishment (RAE), Farnborough, Structures Department, and the document had a NASA reference (N72–28475), with the original RAE reference (RAE-TR-70068) added for good measure. The first words on the cover of the report were 'Ministry of Defence (Aviation Supply)'.

The Royal Aircraft Establishment itself, later to be known as the Royal Aerospace Establishment, with its worldwide reputation for technical excellence and painstaking thoroughness, was a major source of reports, many of which were issued as ARC documents; other RAE items were not available because they were security classified. Inquiries concerning ARC documents should now be addressed to the coordinating body, namely the Defence Evaluation and Research Agency (DERA) Farnborough.

A tidying up operation used to take place in that all ARC *reports and memoranda* were eventually bound into *Annual technical reports volumes*, and as far as possible the bound volume contained the *R&Ms* whose original report dates corresponded with that year, but there were exceptions.

In the academic sphere in Great Britain, a number of institutions are noted for their reports series, particularly the Department of Aeronautics and Astronautics at the University of Southampton, and Cranfield University (previously called the Cranfield College of Aeronautics). A detailed survey of the literature of aerospace engineering has been provided in the second edition of *Information sources in engineering* (MacAdam, 1985), whilst the third edition of the same work treats the topic more diversely: see, for example, the chapter on concurrent engineering (Radcliffe, 1996).

National Aeronautics and Space Administration (NASA)

In the United States, an organization of similar age to the ARC but infinitely larger and more powerful is the National Aeronautics and Space Administration (NASA), which was created in 1958 out of the National Advisory Committee for Aeronautics (NACA) as a civilian agency with the task of accomplishing the United States aeronautics and space programmes. NACA, as noted in Chapter 1, began issuing reports in 1915.

With regard to the provision of information, the National Aeronautics and Space Act of 1958 required that the 'aeronautical and space activities of the United States be so conducted as to contribute to the expansion of human knowledge of phenomena in the atmosphere and space. The Administration shall provide for the widest practicable and appropriate dissemination of information covering its activities and the results thereof'. It is greatly to the credit of NASA that these unequivocal words have been interpreted in the spirit as well as the letter, with the result that a very fine information service has been developed, and to apply the term 'grey literature' to the NASA output seems somewhat inappropriate when the acronym for its announcement service, *STAR* conjures up an image of brightness and clarity.

The NASA Scientific and Technical Information (STI) Program is subject to certain access restrictions, details of which are available (in the case of United States inquirers) from the NASA Center for Aerospace Information, Baltimore; or (in the case of inquirers from outside the United States) from the Director of the STI Program, Washington. A general account of the Program and its role in encouraging greater international cooperation in the aerospace field has been provided by Blado and Cotter (1992), whilst a case for the modernization of the Program has been put by Cotter and co-workers (1992), who suggest that an infrastructure developed by NASA from the mid-1960s to the late 1970s is no longer cost-effective.

NASA and the Department of Defense (DoD) are now jointly engaged in the *Aerospace Knowledge Diffusion Research Project*, concerned with the importance and value of the technical report as a medium for the transfer of federally funded aerospace research and development. The investigation is on-going and progress can be followed in regular updates, as for example NASA TM 107693:1993, and the paper by Pinelli *et al* (1993).

In addition to the announcement journal *STAR*, which is discussed below, NASA disseminates several formal publications series, some of which are as follows:

Contractor reports (NASA CR-) which contain technical information generated in the course of a NASA contract and released under NASA auspices;

Technical memoranda (NASA-TM-X) which provide information receiving limited distribution because of its preliminary or classified nature;

Technical reports (NASA-TR-R) which are documents carrying scientific and technical information considered important, complete and a lasting contribution to existing knowledge;

NASA Technical translations (NASA-TT-F) representing information originally published in a foreign language but considered sufficiently valuable to NASA's work to merit distribution in English.

Other series of a more general kind are:

Conference publications (CP) which are records of the proceedings of scientific and technical symposia and other professional meetings sponsored or co-sponsored by NASA;

Research publications (RP) which are compilations of scientific and technical data deemed to be of continuing reference value;

Special publications (SP) a series concerned with subjects of substantial public interest; they report scientific and technical information derived from NASA programmes and are for audiences of diverse technical backgrounds. An example of a *Special publication* aimed at a general audience is *Records of achievement* (NASA-SP-470) published on the occasion of NASA's 25th anniversary, and including an illustrated narrative on the moon landings and the *Voyager* journeys;

Technical papers (TP) present the results of significant research conducted by NASA scientists and engineers.

As noted above, the predecessor of NASA was NACA, and details of the work carried out under the auspices of the earlier body, a great deal of which is of a fundamental nature and in consequence still referred to, can be obtained by consulting the *Index to NACA technical publications 1915–1949* (1950) and its supplements up to 1960. For a history of NACA and the early days of NASA, see N82–14955 *Orders of magnitude*, covering the period 1915–80. Publications in the *NASA formal reports series* are available as full copy reports through the National Technical Information Service, and cover ten broad areas, namely aeronautics, astronautics, chemistry and materials, engineering, geosciences, life sciences, mathematical and computer sciences, physics, social sciences and space sciences.

European agencies

In continental Europe, several agencies are active in the dissemination of aerospace information. Firstly, there is the internationally composed Advisory Group for Aerospace Research and Development (AGARD), the mission of which is to bring together the leading personalities of the NATO nations in the fields of science and technology relating to aerospace. AGARD distributes unclassified reports and other publications through national distribution centres. The national distribution centre for the United Kingdom is the Defence Information Research Centre, Glasgow, and the UK purchase agency is the British Library Document Supply Centre. A cumulative index to AGARD publications issued since 1952 is regularly updated, and provides information on the major AGARD series, namely:

(1) *Advisory reports (AR)*;
(2) *Reports*;
(3) *Agardographs (AG)*;
(4) *Conference proceedings*.

Full bibliographical references to and abstracts of AGARD publications are provided in *STAR* and in *Government reports announcements and index*. See, for example, N87–29369, abstracts and indexes to AGARD publications issued during the period 1983–85, and prepared with the help of the NASA database. In addition, detailed examples of AGARD publications, and the steps taken to overcome the alleged difficulties in obtaining copies, have been described by Hart (1993).

A CD-ROM product, the *AGARD aerospace database*, has been prepared by W.T.V. GmbH, Karlsruhe and is available from Springer Electronic Media. *The Database* gives details of all publications in the series mentioned above, and covers the years 1960–93. It is a companion compilation to the NATO-PCO (Publication Coordination Office) *Database* covering scientific publications appearing during the years 1973–94. *AGARD reports – a quarterly listing* is available via the NASA-STI home page.

A reference manual which grew out of the AGARD lecture series held in 1988 on the effectiveness of information centres has been compiled by Griffiths and King (1991). The manual considers various aspects of cost-benefit analysis and reviews in particular the value of internal documents (i.e. documents which are not journal articles or books).

In France, where AGARD has its headquarters, the French national agency concerned with aerospace is the Office National d'Etudes et de Recherches Aerospatiales (ONERA), which has published several series of reports and technical notes going back to 1947. In Germany, a similar body responsible for various series of aerospace publications is the Deutsche Forschungs- und Versuchsanstalt für Luft- und Raumfahrt (DFVLR). DFVLR is the largest research establishment for aeronautical engineering in Germany and publishes two major series of reports:

(1) *DFVLR Forschungsberichte (FB series)*;
(2) *DFVLR Mitteilungen (Mitt. series)*.

An index is issued annually. Many of the *FB* and *Mitt.* documents are translated into English by the European Space Agency and published as ESA Technical Translations (ESA TT). In addition to its generally available publications, DFVLR also issues several hundred internal scientific reports each year.

In Europe as a whole, grey literature in the aerospace field is coordinated by the European Space Agency, which as noted above, was created from two earlier bodies, ESRO and ELDO. The ESA members states are Austria, Belgium, Denmark, France, Germany, Ireland, Italy, the Netherlands, Norway, Spain, Sweden, Switzerland and the United Kingdom. The purpose of the Agency is to provide for exclusively peaceful purposes cooperation among European states in space research and technology. ESA has an extensive publications programme with various series of reports in categories similar to those issued by NASA, such as conference proceedings, technical reports, technical memoranda, and so on. ESA processes many items for input to *STAR*.

In addition, the Agency operates a widely used information retrieval service called ESA/IRS, which offers online access to more than 200 databases and databanks in all subjects and disciplines, 20 of which are unique to the service, including EUDISED (see Chapter 10) and NASA. Also unique to ESA/IRS is the *European aerospace database* which covers grey literature, mainly technical reports and conference proceedings acquired by ESA in cooperation with national aerospace organizations and industrial concerns in ESA member states. Coverage includes documents with unrestricted distribution and some with a controlled distribution such as the *ESA Optional Programme Contractor Reports (ESA-CRx)*. Records date back to 1962 and currently the file has over 160 000 references.

ESA/IRS is based in Frascati, Italy, and coordinates a number of ESA/IRS national centres. In the United Kingdom, the national centre is IRS – DIALTECH at the Science Reference Information Service of the British Library, London.

Announcement services

STAR

The aerospace industry has always been well organized in its handling of information, and it possesses a major abstracting and indexing journal covering current worldwide grey literature on the science and technology of space and aeronautics, namely *Scientific and technical aerospace reports (STAR)*. Previous guises of *STAR* were *NASA technical publications announcements* and the *National Advisory Committee for Aeronautics research abstracts*. Publications abstracted in *STAR* cover a large section of grey literature documents and include reports issued by NASA and its contractors, other US government agencies, corporations, universities and research organizations throughout the world. Pertinent theses, translations, NASA owned patents and patent applications are also abstracted.

The subject scope of *STAR* includes all aspects of aeronautics and space research and development, supporting basic and applied research and applications. Aerospace aspects of earth resources, energy development, conservation, oceanography, environmental protection, urban transportation and other topics considered of high national priority in the United States are also covered. *STAR* is arranged by major subdivisions divided into a number of specific subject categories and one general category/division. The major divisions are:

(1) aeronautics;
(2) astronautics;
(3) chemistry and materials;
(4) engineering;
(5) geosciences;
(6) life sciences;
(7) mathematical and computer sciences;
(8) physics;
(9) social sciences;
(10) space sciences.

Despite the careful specification of NASA's subject interests, it has often been noted that *STAR* is full of surprises. Thus, for example, it is possible to find details of a thesis on weekend cottage recreation in Bavaria (N70–26211). Again, NASA's search for truth is not confined entirely to outer space – in the 1970s it participated in the Shroud of Turin Research Project (STURP), using a system of analysis adopted from the Apollo programme. Looking ahead, NASA has published a study called *Quo Vadimus – the 21st century and multimedia* (N92–22663).

Entries in *STAR* are arranged in an unbroken sequence of accession numbers – for example N92–22338, where the N stands for NASA and the digits for the year of accession and the accession number itself. Although the accession number sequence is unbroken, it is not uninterrupted owing to the special analytic treatment given to certain publications such as conference reports or book translations which have separate abstracts written for their component papers or chapters. Citations and abstracts of the component parts follow immediately after those of the parent document. When the subject content of a component paper or chapter differs sufficiently from the general content of the parent, a cross reference citation is also printed at the end of the category to which it would normally have been assigned if it were an independent document.

Five indexes are included in each issue of *STAR*:

(1) subject index;
(2) personal author index;
(3) corporate source index;
(4) contract number index;
(5) accession number index.

Two of these indexes are of special significance. Firstly, as from volume 12, 1974, *STAR* introduced a change with respect to the subject index. Previously, the Notation of Contents (NOC) rather than the title of a document had been used to provide a more exact description of the subject matter. The NOCs were arranged under each broad subject heading in ascending accession number order. Under the new arrangement, NOCs no longer appear in *STAR*'s subject index or in other NASA information products. Instead, titles of the referenced documents are used with a synthesis title extension added in those cases where the actual title is considered insufficiently descriptive. The object, according to NASA, is to enhance the utility and relevance of the subject index. Similar title enhancements are used in other grey literature abstracting publications.

Secondly, the report/accession number index is of particular importance because of NASA's commendable practice of giving full details of a report's original number, that is the one allocated by the issuing agency responsible for the work reported. Report numbers may thus be in the main NASA series, as for example the *Contractor report (CR)*; in series used by NASA's own research centres, such as the *E series* of the Lewis Research Center; or in series used by other organizations both in America and overseas.

The last named category, as noted above, includes reports from RAE. The practical result is that no matter by what number an aerospace document is cited or sought, it is possible to check against the accession number N sequence and so determine its availability. Occasions when such citations occur are typically in personal conversations between scientists and engineers in various establishments and companies, who exchange details of the documents issued by their respective organizations without reference to the NASA system. Subsequently, months or even years later, a library or information department may be asked to supply one of the documents so mentioned, and often the quickest and surest way is to check the N sequence. The report/accession number index provides the means for doing this.

Cumulative index volumes are published semi-annually and annually. The introductory remarks to each issue of *STAR* are particularly well detailed, covering the method of announcement of NASA publications in the journal, the distribution of NASA publications, eligibility and registration for NASA information services and information products, the public availability of documents announced in *STAR* (that is to say, availability within the United States), the European availability of NASA and other documents, and the indexing vocabulary. The introduction also includes examples of typical citations and abstracts and, always providing the points of detail raised are noted, the user will have no difficulty in tracing and obtaining the documents required. With earlier issues of *STAR* up to and including 1971, a sixth index was necessary to relate accession numbers to report numbers since entries were not arranged in one unbroken sequence of N numbers.

The twofold advantage of *STAR*, as stated by NASA and borne out by the experience of many users over the years, is a current awareness service achieved through frequency of publication and a retrospective searching tool arising from thorough indexing in each issue. In fact, it is difficult to overstate the value of *STAR* to any organization intimately or only peripherally concerned with aerospace.

With regard to NASA accession numbers, readers will benefit greatly by familiarizing themselves with one or two limitations concerning the use of the number to secure copies of certain documents. The NASA accession number is sufficient when ordering NASA or NASA-sponsored items, marked with an asterisk (*), from NTIS. When non-NASA documents are wanted (i.e. no asterisk) either direct from the issuing agency, or from other sources, especially NTIS, it is vital that additional background information, including the report number assigned by the originating agency be given. A further symbol commonly used is //, indicating that a document is available on microfiche. Documents which cannot be reproduced as microfiche, but for which one-to-one facsimiles can be provided, are marked with a plus sign (+). Occasionally, report entries may carry no asterisk (i.e. not NASA or NASA-sponsored), no double slash (i.e. no microfiche) and no plus sign (i.e. no one-to-one facsimile), and in these cases availability will depend on an issuing agency's willingness to supply. Documents so listed but difficult to obtain are part of NASA's policy of making *STAR* as comprehensive as possible. Some reports, for various reasons, may not be available directly from their originators, but a different version of the same document may be available from another source. Chillag (1973) quotes the example of N71–17930 which was originally prepared as report NGTE NT 766, but since the paper had been intended for publication, it was not released in that form for general use. The item finally found a home in a volume of *AGARD conference proceedings*, with, however, another NASA number, N71–17372.

The British Library Document Supply Centre maintains an extensive collection of NASA and NASA sponsored publications, but some documents (marked * and/or //) are not always available on demand, even though they may have been announced for some time – usually BLDSC asks the reader to reapply in a few weeks' time, and the interval is used to secure a copy from the United States.

STAR has a companion publication, *Limited scientific and technical aerospace reports (LSTAR)*, which announces reports which are security classified, and unclassified reports with availability restrictions. In 1996, NASA announced that in response to budget restraints and in order to benefit from advances in electronic access, a number of publications would henceforth be available in electronic format only. The titles included *Aerospace medicine and biology, Aeronautical engineering, NASA patent abstracts*, and *Scientific and technical aerospace reports*.

International aerospace abstracts

For many years, the twice monthly issues of *STAR* were complemented by twice monthly issues of *International aerospace abstracts (IAA)*, an abstracting and indexing journal providing worldwide coverage of the open (i.e. non-report) aerospace literature, including scientific and trade journals, books and meetings papers. Bibliographical citations and abstracts were arranged in the same subject categories as those used in *STAR*. Each entry was prefixed by an *IAA* accession number, and each issue of the journal contained indexes for subject, personal author, contract number and accession number. *IAA* is now published monthly and contains English language abstracts of publications originally issued in more than 18 languages from 100 countries. Over 50 per cent of the material originates outside the United States. *IAA* was formerly part of the input to the NASA database, but since 1994 has been used instead to update the *Aeroplus access* database, available via ESA/IRS and STN. *IAA* and the *Aeroplus access* service are produced by the American Institute of Aeronautics and Astronautics (AIAA), and members of AIAA were surveyed (Pinelli, 1991) to determine the most important factors influencing the use by US aerospace engineers and scientists of conference papers, periodical articles, in-house reports and government technical reports. The top three considerations proved to be:

(1) relevance;
(2) technical quality;
(3) accessibility.

Index aeronauticus

An abstracting journal published in Great Britain for over 20 years (the final issue appeared in 1968) and now only of historical interest was *Index aeronauticus*, compiled by the staff of the Ministry of Technology and its predecessors, whose primary purpose was to draw attention to articles of interest in scientific and technical journals, and to published papers and reports within the field of aeronautics. *Index aeronauticus* deserves a mention in any survey of aerospace information sources because of two striking features: its dumpy A5 format, and its use of the Universal Decimal Classification (UDC) for the arrangement of its abstracts.

Retrieval systems

As the reports and other literature amassed by NASA continued to grow, and as the operational need to exploit it to the full became more urgent in the successful pursuance of NASA's programmes, an information retrieval system known as NASA/RECON (from REmote CONsole) was developed and expanded into a national and eventually international network (Jack, 1982). The information retrieval language used was a system referred to as DIALOG, developed by the Lockheed Company and described by Summit (1967), which allowed an important and now familiar facility, the direct interaction between the search requester and the computer in which the data was stored. It provided a means, via a display console, for the user to determine:

(1) what terms were alphabetically related to a specific term or set of terms;
(2) the number of documents in the collection indexed under each displayed term;
(3) terms which were conceptually related to any given term;
(4) what documents dealt with any term or set of terms chosen for searching.

These concepts are familiar enough nowadays, but the Lockheed system was one of the first to offer a large scale online searching service. The link with the Lockheed Company has long been severed, and currently DIALOG Information Services Inc. are a Knight Ridder company.

The NASA Scientific and Technical Information (STI) Program provides a one-stop service embracing unpublished sources (one third of the NASA Aerospace Database) and published sources (two thirds of the Database) covering the following subjects:

Topic	Proportion (per cent)
Engineering	24
Space science	14
Geosciences	11
Chemistry and materials	11
Physics	10
Mathematics and computers	9
Aeronautics	9
Astronautics	9
Life sciences	4
Social sciences	1

(NB rounding results in a total of 102)

NASA has established a partnership with ESA whereby ESA catalogues, abstracts and indexes each document received from its European sources in the NASA format and then provides both paper and microfiche copies with an accompanying electronic citation. NASA loads the new input into the database, merges it with other incoming material and then sends ESA an updated NASA Aerospace Database tape, which ESA in turn incorporates into its own information system. In this way, ESA acts as a consolidator of European input to NASA and as the distributor of NASA information back to the various European institutions. Certain restrictions apply, and access is confined to the ESA member states listed above. NASA publications and other aerospace documents are now available for browsing, downloading and printing via the NSAS STI Home Page and the NASA Center for Aerospace Information Technical Report Server (CASI TRS) using a WAIS search engine. In addition, users can consult the NASA Technology CD-ROM compiled by J. B Data (1996).

Technology transfer

The virtues of technology transfer have become widely appreciated and organizations abound which preach its advantages to industry with missionary zeal. In all this activity, NASA has played a pioneering role, for it recognized that ideas initially developed at great public expense to meet the exacting requirements of the aerospace industry were often capable of modification for application in an extensive range of everyday areas of industry. Consequently, as a special aid to persons and companies not directly involved in aerospace activities, the NASA Office of Technology Utilization began to issue numerous publications regarding advances likely to be capable of transfer from one field to another, and examples of successful spin-off technology began to appear. Moreover, the term technology was interpreted in a very broad sense and not confined to the manufacturing and processing industries. Examples of technology transfer have been reported in various biomedical projects and in public sector initiatives concerned with air and water pollution, housing and urban developments, law enforcement and crime prevention, transport, and fire safety.

Many individual cases of technology transfer have been reported by NASA, and many more are likely, not just from the aerospace sector to industry in general, but also in the reverse direction from industry to aerospace. For example, the *Human exploration initiative* (NASA, 1990) reports on a variety of approaches to lunar and Mars exploration and explains the resources and strategies likely to be required, highlighting the contribution of ideas which can be made by universities, research establishments and industrial companies. The papers presented at *Technology 2001: the Second National Technology Transfer Conference and Exposition* (N92–22423), an event organized by NASA, cover advanced manufacturing, artificial intelligence, biotechnology, computer graphics and simulation, communications, data and information management, electronics, electro-optics, environmental technology, life sciences, materials science, medical advances, robotics, software engineering, and testing and measurement.

The main NASA media for announcing possible technology transfer opportunities are *Tech briefs, Technology utilization reports* and *Technology utilization surveys. Tech briefs* are short announcements of new technology derived from research and development activities carried out by NASA. They emphasize information likely to be transferable across industrial, regional and disciplinary lines, and are issued to encourage commercial applications. *Technology utilization reports* are of a far more substantial nature, and give detailed descriptions of developments and innovations of promise. The position is similar with *Technology utilization surveys*, which are comprehensive state-of-the-art accounts identifying substantial contributions to technology by NASA researchers or NASA contractors. Both types of documents appear in the NASA *SP series*, noted above. NASA also produces *Research and technology objectives and plans (RTOPS)* which are summarized to facilitate communication and coordination among interested technical personnel in government, industry and the universities (see, for example N88–14894). Specific issues are also reported in depth, as, for example the use of expert system in technology transfer (N88–14882). Technology transfer as an identified activity has spread well beyond the aerospace industry. Knox (1973) in a survey of systems for technological information transfer, noted in particular the efforts of the Department of Defense, the Atomic Energy Commission and the Office of Education; he pointed out that in each case one of the distinguishing features was the emphasis on the technical report. Technology transfer also takes place by osmosis and diffusion, and industry has not been slow to exploit opportunities as they arise. Indeed, it has been argued that one of the by-products of the programme which culminated in the American moon landings has been the development by a whole range of companies of smaller and more reliable lightweight computers.

Summary

The grey literature of the world of aerospace, as with other subject areas, is characterized by a diversity of form and content. Hundreds of agencies contribute many thousands of documents each year to a common pool of information and data. Consistent and comprehensive efforts are made to ensure that as many publications as possible reach a wide and diffuse readership at the earliest opportunity, subject only to the constraints of security restrictions and the need-to-know.

References

Blado, W.R. and Cotter, G.A. (1992) An international aerospace information system: a cooperative opportunity. *Online review*, **16**, (6) 359–368

Chillag, J.P. (1973) Don't be afraid of reports. *BLL review*, **1**, (2) 39–51

Cotter, G.A., Hunter, J. and Ostergaard, K. (1992) Modernization of the NASA Scientific and Technical Information Program. *Online Information 92: Proceedings of the 16th International Online Information Meeting.* London, (ed.) D.L. Raitt. Oxford: Learned Information (Europe) Ltd, 273–279

Griffiths, J.M. and King, D.W. (1991) *A manual on the evaluation of information centers and services.* New York: American Institute of Aeronautics and Astronautics

Hart, D.W. (1993) Aerospace and other grey literature from NATO. In Farace, D.J. (ed.) (1994) *Proceedings of the First International Conference on Grey Literature, Amsterdam, 1993.* Amsterdam: TransAtlantic Publishing

J.B. Data (1996) CD-ROM holds NASA technology. *Information today*, **13**, (2) 25

Jack, R.F. (1982) The NASA Recon search system. *Online*, 6 November, 40–54

Knox, W.T. (1973) Systems for technological information transfer. *Science*, **181**, (4098) 415–419

MacAdam, E.J. (1985) *Aerospace engineering.* In Anthony, L.J. (ed.) *Information sources in engineering,* 2nd edn. London: Butterworths

National Aeronautics and Space Administration (1989) *NASA information system: its scope and coverage.* N89–15779

Pinelli, T.E. *et al* (1991) The relationship between seven variables and the use of US government technical reports by US aerospace engineers and scientists. In Griffiths, J.M. (ed.) *ASIS '91: systems understanding people.* Medford, N.J.: Learned Information Inc., 313–321

Pinelli, T.E. *et al* (1993) NASA/DoD Aerospace Knowledge Diffusion Research Project, paper 35: the US government technical report and aerospace knowledge diffusion – results of an ongoing investigation. In Farace, D.J. (ed.) (1994) *Proceedings of the First International Conference on Grey Literature, Amsterdam, 1993.* Amsterdam: TransAtlantic Publishing

Radcliffe, D.F. (1996) Concurrent engineering. In Mildren, K. and Hicks, P. (eds.) *Information sources in engineering*, 3rd edn. London: Bowker-Saur

Summit, R.K. (1967) Dialog: an operational online reference retrieval system. *Proceedings 22nd National Conference, Association of Computing Machinery*, 51–56

CHAPTER SEVEN

Life Sciences

Introduction

The definition of the term life sciences depends on which authority is consulted. A dictionary will say a life science is any of the sciences (such as zoology, bacteriology or sociology) which deal with living organisms; or such sciences collectively. In the case of literature databases, the life sciences collection provided by PASCAL offers a convenient way of defining the area under consideration, since it embraces biology, medicine, psychology, pharmacology and agronomy. Another approach is that used by the BIOSIS Information System, which says the life sciences relate to the study of living things, that is animals (including humans), plants and micro-organisms with respect to:

(1) how they look (anatomy, morphology, cytology);
(2) how they live (physiology, biochemistry);
(3) how they relate to the environment (ecology, behaviour);
(4) how they grow(reproduction, evolution, genesis);
(5) how they are identified (taxonomy, systematics).

For the purposes of this chapter, these various categories have been consolidated into the headings medicine and biology, and agriculture and food.

The conventional literature relating to the life sciences has been considered separately and comprehensively in other titles in this series, especially Morton and Godbolt's *Information sources in medical sciences* 4th edition (1992); Lilley's *Information sources in agriculture and food science* (1981); Wyatt's *Information sources in the life sciences*, 4th edition (1997); Pickering's *Information sources in pharmaceuticals* (1990); and Wood, Hardy and Harvey's *Information sources in the earth sciences*, 2nd edition (1989). None of these titles highlights reports in particular or grey literature in general as being of special significance, and this is to some extent explained by the absence of an announcement journal devoted exclusively to grey literature in the life sciences. The title which perhaps comes nearest is *Biological abstracts/RRM (Reports, reviews, meetings)* which complements *Biological abstracts*, easily the most comprehensive source of life sciences journal literature. *Biological abstracts/RRM* adds approximately 215 000 records each year and contains nearly 2 million records on a CD-ROM disc embracing the years 1989 to date. The online version of the two publications is available from BIOSIS, a non-profit organization based in Philadelphia whose origins go back to 1926.

In Europe, the German Institute for Medical Documentation and Information (Deutsches Institut für Medizinische Dokumentation und Information (DIMDI), founded in 1969 with headquarters in Cologne, offers access to over 80 databases in the life sciences and the social sciences.

Medicine and biology

On the question of grey literature in medicine and biology, an important factor noted by Morton and Godbolt (*op.cit.*) is that 'the indifference which most scientists feel for the research report literature is due not so much to a low opinion of reports as to a respect for the traditional primacy of the 'open' literature'. It has also been noted that as a form of primary communication, reports are of much less significance in medicine and biology than in other areas of science and technology, and a major reason for this state of affairs is that there is far less government controlled research and development than in, say, space projects or nuclear energy activities. Even when state bodies are involved, such as the Medical Research Council and the American Public Health Service, their research reports are published not as communications directly emanating from a research establishment with all the characteristic identifying features of a typical report document, but as conventional publications of the respective government publishing systems.

The question of reports and their treatment in medical libraries has been considered by Sargeant (1969), who notes that many reports lead directly to journal articles A more recent study by Goldschmidt (1991) looks at citation counting as a means of assessing the quality of research reports, using the example of documents in corticosteroid therapy for bacterial infections. Indeed, there is a growing amount of grey literature in medicine and biology, sufficient in fact to warrant special surveys, as for example the study by Albarani and co-workers in Italy (1990) and by Otake (1990) in Japan. In the specific area of pharmaceuticals, Hirvensalo (1992) has described the value of grey literature as a source of commercial intelligence, whilst the selection of grey literature produced by the nursing profession has been examined by Cianelli and Leyton (1995).

The breakdown of barriers between scientific disciplines has led to the appearance of more and more biological and medical information being cited in the major grey literature announcement services, and the extent to which topics now feature is shown by a comparison of four key publications. Firstly, *Government reports announcements and index (GRA&I)*, published by the National Technical Information Service (NTIS) has three major medical and biological headings in the NTIS subject category and subcategory structure, namely:

(1) Biomedical technology and human factors engineering, with subcategories in instrumentation, bionics and artificial intelligence, life support systems, protective equipment, prosthetics and mechanical organs, and tissue preservation and storage;
(2) Health planning and health service research, with subcategories in a wide range of health care activities and administration;
(3) Medicine and biology, with subcategories in anatomy, biochemistry, botany, clinical chemistry, clinical medicine, cytology, genetics and molecular biology, dentistry, ecology, electrophysiology, immunology, microbiology, nutrition, occupational therapy, parasitology, pest control, pharmacology, physiology, psychiatry, public health, radiobiology, stress physiology, surgery, toxicology and zoology.

This extremely comprehensive list of subjects serves to emphasis *GRA&I*'s role as a major source of items in the medical and biological fields. The second announcement service to be considered, *Scientific and technical aerospace reports (STAR)*, published by the National Aeronautics and Space Administration (NASA) uses the following headings:

51 Life sciences (general)
52 Aerospace medicine, including physiological factors, biological effects of radiation, and the effects of weightlessness on man and animals;
53 Behavioural sciences, including psychological factors, individual and group behaviour, crew training and evaluation, and psychiatric research;
54 Man/system technology and life support, including human engineering, biotechnology, and space suits and protective clothing;
55 Space biology, including exobiology, planetary biology and extraterrestrial life.

The headings indicate an applications-oriented coverage and form the basis of a growing specialization in the life sciences, aerospace, medicine and biology. The publications in these areas have expanded to such an extent that NASA finds it appropriate to publish a continuing bibliography, which appears in the *Special publications* series. The bibliographies give details of NASA's range of programmes, as for example *Biomedical results from Skylab* (N77–33780) and *Space gerontology* (N83–16018). Other titles from *STAR* in the life sciences include *Control of blood pressure in humans under micro-gravity* (N92–23071) and *Acoustically based fetal heart rate monitor* (N92–22733).

The third announcement service which needs to be considered with respect to entries in the life sciences is *Energy research abstracts (ERA)* from the Office of Scientific and Technical Information of the United States Department of Energy (DOE). The abstracts on medicine and biology are arranged under three main headings:

55 Biomedical sciences, basic studies, including behavioural biology, biochemistry, cytology, genetics, metabolism, medicine, microbiology, morphology, pathology, physiological systems, and public health;
56 Biomedical sciences, applied studies, including radiation effects, thermal effects, chemical metabolism and toxicology, and other environmental pollutant effects;
57 Health and safety (no subcategories).

In addition, various environmental issues are covered under a further four headings. Many of the life sciences items noted in *ERA* refer, as befits the successor to *Nuclear science abstracts*, to radiation and environmental problems, as for example *The shape of the cancer mortality dose-response curve for atomic bomb survivors* (90–33229) and *Health and safety guide for inorganic compounds and metals used in the fabrication of superconductive alloys* (90–33234).

Finally, the fourth journal to turn to when seeking life sciences material is *Resources in education (RIE)* from the United States Department of Education, which, whilst it has no obvious medical and biological headings among the sixteen Clearinghouse sections by which it is arranged, does carry a considerable number of references to life science topics, as a check of the subject index will readily reveal. Typical entries include *Asthma in children* (ED 340155), *Health education* (ED 340680) and *Coping with stress* (ED 339 794).

The foregoing announcement services are all American in origin; in the United Kingdom, items in the grey literature concerned with biology and medicine are monitored by the staff at the British Library Document Supply Centre and announced in *British reports translations and theses (BRTT)*, using the modified COSATI classification scheme, section 06 of which (Biological and medical sciences) has nearly 30 subdivisions. The topics covered include biochemistry, bio-engineering, human anatomy, clinical medicine, hygiene and sanitation, pharmacology, human psychology, wounds and injuries, botany and zoology. Many of the entries under headings such as biochemistry and clinical medicine relate to PhD theses, whilst those under the heading industrial (occupational) medicine carry details of translations originating with the Health and Safety Executive, London. Typical entries are *Monitoring cataract change* 1990 (PhD thesis, DSC D96778), *Stress in the dealing room* 1990 (PhD thesis, DSC D96809), *Autistic spectrum disorders* 1995 (National Autistic Society, DSC:96/24448) and *Catheters for intermittent self-catheterization* 1996 (Medical Devices Agency, DSC:3595.420480(A18)).

In addition, the British Library's current awareness bulletin *Focus on British biological and medical sciences research* covers the latest reports added to the National Reports Collection, plus translations and theses. The German grey literature journal *Forschungsberichte* covers biology and medicine under section 06, and the items listed are available from the TIB Hanover.

Diversity of interests

Grey literature in medicine and biology ranges from the highly specialized (e.g. *Genetics-based machine learning and behaviour-based robotics*, N92–22419) to matters of public concern (e.g. *Volunteer lake monitoring: a methods manual for lake water quality measurement by citizen volunteers* PB 92–218 411). Many publications are announced and discussed in the general press, especially those from pressure groups, charities and specialized institutions dealing with issues at the uneasy interface between professionals and enlightened laymen. Titles representative of this type of grey literature include the *Alzheimer's disease report* (the Alzheimer's Disease Society, 1992), *Life will never be the same again* (on falling attendances at ante-natal clinics), (the Health Education Authority, 1992), and *Hungry in hospital?* (on allegations about underfed patients in a study compiled by the Association Of Community Health Councils, 1997).

The diversity of such material, the amount of which shows no signs of abating, emphasizes more than ever the need for proper bibliographical control and ease of access. Some materials indeed are adequately covered in standard reference sources, provided the searcher remembers to check them. For instance, a great number of biological and medical publications originate in government departments, and are recorded in Chadwyck-Healey's *Catalogue of British official publications not published by HMSO*, which may not be the obvious source for details of *Biotechnology in Japan* (Science Engineering Research Council) or *International response to drug misuse* (Foreign and

Commonwealth Office). The *catalogue* is, as noted in Chapter 1, now issued in a CD-ROM format which combines HMSO/The Stationery Office records with those from Chadwyck-Healey, under the new title *UKOP: catalogue of United Kingdom Official Publications.*

Finally, medicine provides an excellent example of the working of the United States Freedom of Information Act referred to in Chapter 2, namely the ability to access unpublished reports and other information contained in the files of the US Food and Drugs Administration (FDA). The FDA is responsible for ensuring the safety and efficiency of all pharmaceuticals and medical devices introduced into interstate commerce, and in order to discharge this responsibility, reviews approval applications for new products, tracks product performance after approval, inspects manufacturing facilities, and provides guidance for manufacturers in complying with regulations. The information accumulated as a result of these activities is naturally of great public interest, and the DIOGENES database (a joint venture of FOI Services Inc. and Washington Business Information Inc.) includes details of unpublished FDA documents acquired under the Act. DIOGENES contains over 630 000 records, including the full texts of thousands of FDA documents. Details are given in the *DIOGENES users' guide*, available free on request.

Agriculture and food

The apparent reluctance of scientists in medicine and biology to use reports as a medium for disclosing the results of their work is not shared by their colleagues in agriculture and food. For almost two hundred years, accounts of research and development and advisory or extension work have been published for the benefit of all kinds of workers in the primary British industry – agriculture. Originally, publication was in the journals and papers of the various agricultural societies (for example *Bath and West 1877-; Royal Agricultural Society, 1840-)* and for nearly one hundred years the results of work have also appeared as separate monographs and reports.

Depending on the stage of development in a country, food production is more or less closely linked with agriculture – at subsistence level, virtually one and the same activity; at a highly sophisticated agricultural level, they are quite separate. In the literature, apart from food technology in the extreme, there is quite a degree of overlap, and differentiation between the two areas will be avoided here.

There is a considerable wealth of grey literature available in agriculture and food, and it has been well represented in the information explosion as progressively more funds and personnel have been employed in research and development. Publications pour out from a wide variety of bodies: from the Food and Agriculture Organisation (FAO) of the United Nations; from the Organisation for Economic Cooperation and Development (OECD); from government and university departments of agriculture; from agricultural research bodies; and from research stations general in scope or devoted to one or more component disciplines of the subject field.

Because the subject content of agriculture is broadly applied biology, the pattern of the production and publication of reports and other items of grey literature is worldwide on a zonal or regional basis rather than purely national in character. Climate, geological conditions, vegetation, pests and diseases overrun national boundaries and contribute to the broad picture of cultivation and consumption. Nevertheless, these topics are so closely concerned with national economies and national prosperity that there is considerable official activity and involvement, and publication in many countries tends to follow a common route. As a result, most of the material is eventually recorded in national bibliographies, catalogues of most governments' official publications and, to a lesser extent, in commercially produced bibliographies.

The whole of the literature of agriculture (both conventional and grey) is an area which has been subject to intense and extensive activity. Not only are there excellent guides such as Lilley's *Information sources in agriculture*(noted above), outstanding international announcement services such as the *Bibliography of agriculture*, and specific reference aids like Bush's *Agriculture: a bibliographical guide* (1974); there is also the International Association of Agricultural Librarians and Documentalists (IAALD). The IAALD has looked at grey literature in food and agriculture in considerable depth, and an overview by Chillag (1982) chronicles the attempts which have been made to measure the extent of the material in terms of numbers of publications and sources, and to come to an agreement on how it should be handled. Chillag quotes Burntrock (working in the 1970s) as referring to the existence of over 500 secondary announcement services in agriculture, and that on average the same document may be abstracted in seven different publications.

The solution to this duplication of effort, Chillag suggests, may lie in AGRIS, the International Information System for the Agricultural Sciences and Technology, which was established following recommendations at the 1971 conference of the Food and Agriculture Organisation. AGRIS provides information on agriculture from an input supplied by over one hundred national centres to the co-ordinating centre in Rome. Types of literature covered include journal articles, conference proceedings, monographs, theses, patents, maps, technical reports, standards, films and computer files. All branches of agriculture are included in the material supplied to the AGRIS database, which is also available in paper format as *AGRINDEX*. It is the intention of

AGRIS to improve the input of non-conventional material, and the organizers are actively acquiring publications from Third World countries and from eastern Europe. However, the system faces formidable competition from *CAB* and *AGRICOLA*, noted below.

In the United States, a survey (Smith, 1991) to discover the methods which Land Grant Libraries used to acquire, process and maintain publications emanating from the State Agricultural Experiment Stations (SAES) and the US Department of Agriculture (USDA) found that the most frequently mentioned reference tools were *AGRICOLA/ Bibliography of agriculture*, the *Monthly catalog of government publications*, and *Government reports announcements and index*. In the Caribbean Community (CARICOM), grey literature relating to agriculture has come under scrutiny from Bandara (1987), whilst in Africa the role of unpublished reports in meeting the needs of agricultural researchers has been described by Kaniki (1992).

Raju(1991) has compared three major abstracting services, namely *Rural development abstracts, International development abstracts* and *Abstracts of rural developments in the tropics* to examine their scope and pattern of citation with respect to over 6000 abstracts published during the period 1986–88, which pertain to eleven Asian countries, and the point is made that coverage of grey literature is inadequate. The author suggests that publishers might make an effort to include more grey literature items by cooperating with regional and international agencies.

A detailed study of the grey literature of agricultural economics (Cesare and Sala, 1995) has introduced a degree of shading in that the terms dark, medium and light grey are used to account for different levels of availability and access. This refinement of different categories of grey is applied equally to reports, theses and preprints.

Announcement services

The role of announcement services in food and agriculture will be considered in the same sequence as that used for medicine and biology. Firstly, there is *Government reports announcements and index (GRA&I)*, already noted as an important source of information on reports in agriculture. The NTIS subject category and subcategory system uses three main headings in this area, as follows:

(1) Agriculture and food, with subcategories in agricultural chemistry; agricultural economics; agricultural equipment, facilities and operations; agricultural resource surveys; agronomy, horticulture and plant pathology; animal husbandry and veterinary medicine; fisheries and aquaculture; and food technology;

(2) Environmental pollution and control, with a subcategory on pesticide pollution and control;

(3) Natural resources and earth sciences, with subcategories on forestry and soil sciences.

Typical entries include the *Saiya (Kenya) sugar project feasibility study* (PB 92–201 623) and *Pesticide toxicity in freshwater fish* (PB 92–858 414).

In *Scientific and technical aerospace reports (STAR)*, coverage of agriculture and food is, as one would expect, somewhat restricted. The headings allocated by NASA vary, and it is necessary to look under a number of places, as the following examples reveal: *Application of Marine Observation Satellite (MOS-1) data for skipjack fishing ground forecast* (N92–22906) is found under category 43 'Earth resources and remote sensing', and *Forestry: environmental aspects* (N92–22847) is found under category 42 'Geosciences' (general). The Department of Energy's *Energy research abstracts (ERA)* conceals the heading 'agriculture and food technology' as the final subsection of category 55: Biomedical sciences – basic studies. References are scattered throughout the publication and include many on the effects of acid rain. Specific topics range from *Solar energy for a hot air dryer at the Lie Farmstead in the County of Flesberg* (13:14866) to *Radioactivity: a layman's guide*, issued by the British Food Manufacturing Industries Research Association (13:16947). *Resources in Education (RIE)* has a sprinkling of references to food and agriculture, as for instance *Homestead and gardening skills* (ED 339 811), *Nutrition instruction* (ED 339 889) and *Wildlife management* (ED 340 534).

In Great Britain, *British reports translations and theses (BRTT)* lists items under subject categories thus:

02 Agriculture, plant and veterinary sciences, which covers agricultural chemistry, agricultural economics, agricultural engineering, agronomy and horticulture, animal husbandry, forestry, veterinary sciences, and fisheries and aquaculture;
06 Biological and medical sciences, which covers food technology.

Documents on agriculture and food are well represented in *BRTT*, and typical entries include *Directory of pear cultivars*, Agroforestry Research Trust (DSC:q96/22565); *Report on the welfare of farmed fish*, Farm Animal Welfare Council (DSC:GPC/09583), and *Studies on the Bacillus flora of milk and milk products*, Glasgow Caledonian University PhD Thesis (DSC:DX191622).

In Germany, *Forschungsberichte* announces items of agricultural interest under the now familiar heading 02 Agriculture, plant and veterinary sciences, and as with reports on medicine, items listed are available from the TIB Hanover.

All the major bibliographical services dealing with the conventional literature in food and agriculture do also include a certain amount of grey literature material. For example, the International Food Information Service (IFIS) provides printed and electronic versions of *Food science and technology abstracts (FSTA)* which contain information from around 1600 scientific journals as well as selected books, proceedings, reports, patents and items of legislation. IFIS was founded in 1968 and is governed by a Board of Governors of eight persons, two representing each of the four sponsors, namely CAB International (CABI), Zentralstelle fur Agrardokumentation und Information (ZADI), the Institute of Food Technologists (IFT), and the Centrum voor Landbouwpublikaties en Landbouwdocumentatie (PUDOC). IFIS has help desks in Great Britain (Reading), North America (Riegelsville, PA), Malaysia (Kuala Lumpur) and Germany (Frankfurt am Main).

CAB International (once known as the Commonwealth Agricultural Bureaux) publishes a range of abstract journals in printed and electronic form covering topics such as animal nutrition, ecology, genetics, leisure and tourism, plant breeding and veterinary science. The origins of the abstracts go back to 1912 when items on entomology were first prepared. *CAB abstracts* covers over 11 000 serial sources as well as conference proceedings, books, reports patents and published theses. The service is available online and in CD ROM format.

Underpinning the whole CAB activity is the *CAB thesaurus*, first published in 1983. It represents the world's largest English language thesaurus for the agricultural and related sciences, and is used for the in-depth indexing of the *CAB abstracts* series, and also AGRICOLA, the database of the US National Agricultural Library (NAL). AGRICOLA (Agricultural Online Access) is a family of data files with indexes to world journals and monographic literature and to United States technical reports on agriculture, agricultural economics, food and nutrition, and related topics.

AGRICOLA, formerly known as CAIN, consists of a number of subfiles, including indexed material from the American Agricultural Economics Documentation Center, the Food and Nutrition Education and Information Materials Center, and several other specialized agencies. The printed version of AGRICOLA is *Bibliography of agriculture*, an announcement service dating back to 1942, and the file, like many other databases today, is also available on CD-ROM (from 1970).

Other sources of information

As with biology and medicine, it is possible to find references to grey literature on agriculture and food in a variety of information sources. Certainly, recourse should be made to Chadwyck-Healey's *Catalogue of British official publications not published by HMSO*, (now called *UKOP* – see Chapter 1), in order to identify such titles as *Developments in food making in the USA* (Department of Agriculture for Northern

Ireland) and *Protected landscapes* (The Countryside Commission). Matters relating to food and agriculture are of great interest, not to say concern, to the public at large, and many grey literature documents dealing with questions of the day are regularly noted in the general press. Typical of such items are the *Report on farming in the Eastern counties of England* (Department of Land Economy, Cambridge 1992), which reflected on the boost sterling devaluation had given to farm incomes, and the document *Red meat* (Meat Hygiene Service) which dealt with the *E. coli* bacterium, and was completed as a 54-page draft in December 1995. A controversial amended version was made available to the public in August 1996 (Newton and Hornsby, 1997).

Summary

Existing announcement and abstracting arrangements for conventionally published life sciences literature have reached a mature stage of development, and the coverage seems to be more than adequate. It is possible to access life sciences databases via a single source by using the Deutsches Institut für Medizinische Dokumentation and Information (DIMDI), which after initially concentrating on health and medicine, has widened its scope to cover the entire spectrum of the life sciences and many areas of the social sciences as well. In all, DIMDI provides access to over 50 databases with a total of more than 40 million records.

Grey literature in these areas does not seem to be so satisfactorily covered. For example, in agriculture, only relatively few grey literature document entries can be found in *AGRINDEX*, and even these can be attributed to just a few input centres. Formal notification of life sciences grey literature has a long way to go, compared with what has been achieved in energy or aerospace, both of which, as has been noted above, have many links with the life sciences. It is true, of course, that in energy and aerospace non-conventional documents assume major proportions, but nevertheless there are useful lessons to be learned on how to handle them. In the case of *Scientific and technical aerospace reports (STAR)* access to documents can be gained centrally via the National Aeronautics and Space Administration, whilst at the International Nuclear Information System (INIS) (see Chapter 11), the flow of documents is derived from national centres.

There are advantages in processing input to such databases at access points through which very large numbers of non-conventional items pass. A lot of fringe material will inevitably be missed by centres which collect in narrow subject fields only. What seems to be of prime importance is that life sciences databases are not starved of grey literature just because it is thought that such input requires a special sort of expertise. The procedures used by organizations in nuclear energy and aerospace have already been mentioned, and this chapter has sought to demonstrate that their respective announcement services have items of great relevance to workers in the life sciences.

A choice lies between the sectorial approach, that is to say the sending of details of grey literature to the most appropriate database in the life sciences, or the general approach, namely the sending of details to a grey literature collection point (in the case of Great Britain, the British Library Document Supply Centre), firstly for publication in *British reports translations and theses*, and secondly for incorporation in the multinational database *SIGLE*. At present, both avenues are used.

References

Albarani, V. *et al* (1990) The use of grey literature in the health sciences – a preliminary survey. *Bulletin of the Medical Library Association*, **78**, (4) 358–363

Bandara, S.B. (1987) Grey literature in the CARICOM region. *Revista AIBDA*, **8**, (1) 1–14

Bush, E.A.R. (1974) *Agriculture: a bibliographical guide*. London: Macdonald and Janes

Cesare, R.D. and Sala, C. (1995) The use of grey literature in the agricultural economics field: a quantitative analysis. In Farace, D.J. (ed.) (1996) *Proceedings of the Second International Conference on Grey Literature, Washington, 1995*. Amsterdam: TransAtlantic Publishing

Chillag, J.P. (1982) Non-conventional literature in agriculture – an overview. *IAALD quarterly bulletin*, **27**, (1) 2–7

Cianelli, R. and Leyton, V. (1995) Criteria for selecting grey literature. In Farace, D.J. (ed.) (1996) *Proceedings of the Second International Conference on Grey Literature, Washington, 1995*. Amsterdam: TransAtlantic Publishing

Goldschmidt, P.G. (1991) Are higher quality medical studies quoted more often? *Health information and libraries*, **2**, (3) 128–139

Hirvensalo, U. (1992) Commercial intelligence through grey literature *Proceedings 2nd Annual European Business Information Conference*, Brussels. London: Task Force Pro Libra, 173–184

Kaniki, A.M. (1992) Meeting the needs of agricultural researchers in Africa: the role of unpublished reports. *Information development*, **8**, (2) 83–89

Lilley, G.P. (ed.) (1981) *Information sources in agriculture and food science*. London: Butterworths.

Morton, L.T. and Godbolt, S. (eds.) (1992) *Information sources in the medical sciences*, 4th edn. London: Bowker-Saur

Newton, P. and Hornsby, M. (1997) Ministers in conflict over report on meat hygiene. *The Times*, 7 March 1977

Otake, H. (1990) In-house searching and collection of government information – case study on materials from Japan's Ministry of Health and Welfare. *Pharmaceutical library bulletin*, **35**, (4) 211–227

Pickering, W.R. (ed.) (1990) *Information sources in pharmaceuticals*. London: Bowker-Saur

Raju, K.A. (1991) Rural development literature: preliminary results of a comparative study of three global abstracting services. *Quarterly bulletin of the International Association of Agricultural Librarians and Information Specialists* **36**, (4) 227–232

Sargeant, C. W. (1969) Handling technical reports in the medical library. *Bulletin of the Medical Library Association*, **57**, (1) 41–46

Smith, H. (1991) Agricultural documents: acquisition and control. *Special libraries*, **82**, (1) Winter 23–29

Wood, D.N. *et al* (eds.) (1989) *Information sources in the earth sciences*, 2nd edn. London: Bowker-Saur

Wyatt, H.V. (ed.) (1997) *Information sources in the life sciences*, 4th edn. London: Bowker-Saur

CHAPTER EIGHT

Business and Economics

Introduction

At the general level, economic and business studies, although they are often classed as part of the social sciences, have much in common with science and technology in the structure of the literature and the way in which it is used. In particular, periodical articles have increased in importance, and in recent years working papers, essentially unpublished drafts of potential periodical articles, have achieved a more important role. The general literature in this area has been covered very thoroughly by three other works in this series, namely *Information sources in economics* (Fletcher, 1984), *Information sources in management and business* (Vernon, 1984), and *Information sources in finance and banking* (Lester, 1995), and in consequence the present chapter will confine its attention to the grey literature as represented by the following categories of publications:

(1) theses;
(2) research and development reports;
(3) working papers;
(4) market research reports;
(5) reports from banks and stockbrokers.

It is worth pointing out that the problems of disseminating information about, tracing the origins of, and obtaining copies of items of grey literature are relatively acute in the social sciences, partly because the material is of fairly recent growth, and partly because so far there are few bibliographical tools devoted specifically to it.

Theses

The main bibliographical works for theses, already described in Chapter 5, give comprehensive coverage to economics and business subjects, and it is sufficient here to note the special features of each in the treatment of the social sciences area.

The major source of information is *Dissertation abstracts international (DAI)*, section A of which deals with the humanities and social sciences, and subsection 5 specifically covers the social sciences in North America. For the coverage of European theses, the reader must consult Section C. A closer look at this subsection reveals that among other subjects it treats business administration (embracing accounting, banking, management and marketing) and economics (dealing with agriculture, commerce-business, finance, history, labour and theory). *DAI* is published by University Microfilms International, which also issues *American doctoral dissertations (ADD)* and *Masters abstracts (MAI)*, both of which compilations should be consulted for data on business and economics.

Theses originating in Great Britain and Ireland are recorded in the Aslib *Index* and the heading for economics has a number of subdivisions. More specifically, it is possible to trace business and economics theses through a number of journals, notably the December issues of *American economic review*; the British title *Economic journal;* and the March issues of the *Journal of economic history*, which carry extended abstracts of recently submitted economic history theses.

Research and development reports

The bulk of government expenditure for research and development tends to be in the scientific and technical areas, but nevertheless the major announcement journals, including *Government reports announcements and index (GRA&I), Scientific and technical aerospace reports (STAR)* and *selected RAND abstracts (SRA)* each have a section devoted to business and economics. This is not particularly surprising in the case of *GRA&I*, since the National Technical Information Service (NTIS) is an agency of the US Department of Commerce.

The relevant section in *GRA&I* is, under the old COSATI system, field 05: behavioural and social sciences, and under the newer NTIS subject category and sub-category structure: administration and management, and business and economics. Two related headings are problem-solving information for state and local governments; and urban and regional technology and development. Documents cited range from reports such as the Central Intelligence Agency's *Eastern Europe: struggling to stay on the reform track* (PB92 928 021) to the many export trade information compilations and fact sheets prepared by the International Trade Administration, Washington.

In *STAR*, heading 83 is devoted to economics and cost analysis, but broader issues can appear under different headings. For example, the *Proceedings of the second symposium on space industrialization* (N85–11–011) discuss economic factors and space commercialization, but are entered under Astronautics (general). Reports issued by RAND, the independent non-profit organization engaged in scientific research and analysis, include the *Financial cost of export credit guarantee programs* (R-3491) and *Cost based reimbursement for nursing home care* (P-7353-RGS).

For reports on business and economics issued in the United Kingdom, the prime source of information about new titles is the British Library's *British reports translations and theses (BRTT)*. The arrangement of the entries follows the modified COSATI scheme mentioned in Chapter 12, and sections 05A: Management, administration and business studies, and 05D: Economics and economic history, give details of documents originating in a wide variety of organizations, including universities, foundations, firms of consultants, and charities. The subject matter extends from *The profitability of horse racing* (88–08–05D-002) to the *Determinants of fixed capital investment* (88–08–05D-041), and from *Exchange rate regimes and the persistence of inflation* (92–09–05D-010) to *Retail rents and market areas* (9209–05D-041), and *Contagious currency crises* (96–11–05D-059). The British Library's current awareness bulletin *Focus on British business and management sciences research* covers the latest reports added to the National Reports Collection, plus translations and theses.

Government funded research and development in the business and economics areas is channelled through a number of agencies, one of the most important of which is the Economic and Social Science Research Council (ESRC). The ESRC, which was established by Royal Charter in 1965, supports research carried out in universities and research centres in Britain into economic life and social behaviour. Its reporting requirements specify that end-of-award reports, preferably in the form of succinct documents of not more than 5000 words in length, shall be deposited in the British Library Document Supply Centre, and thus made available for public borrowing. In addition, the award holders authorize the Council and BLDSC to disseminate reports by copying, microfilm, microfiche or other suitable means.

This procedure represents a definite effort to ensure that the ESRC's research results are clearly directed along the correct routes for inclusion in the grey literature monitoring services and databases. In addition, of course, announcements are made in the ESRC's own publications, such as *Research supported by the ESRC* and the *ESRC newsletter*

Working papers

The working paper or discussion paper has acquired a special significance in business and economics, and Fletcher (1984) has memorably recounted how one eminent academic described the sequence of publication as 'First a draft paper circulated to a small select group of colleagues upon whose discretion he could rely if the paper was bad; second a revised draft duplicated in sufficient quantities (usually a few hundred) to send to interested individuals and organizations; and finally a manuscript submitted for publication to a learned journal'.

Such a sequence takes time, especially if due allowance is made for colleagues to record their comments, and the result is a considerable body of information in circulation in a less than well organized manner. However, thanks to a commendable initiative by the University of Warwick, which recognized the importance of economics working papers and the problems which they created for research workers and librarians alike, a collection of such material was started at Coventry in 1968. The collection contains working papers in economics and management, and from other areas where material is related to economics and management, such as demography, statistics, development studies, and law. Working papers are received in the library from the university itself, from other universities and colleges, business schools, research institutions, banks, and government and other official organizations. For many years, working papers have been listed in bibliographies produced by the library staff at Coventry and then published in conjunction with the Trans-Media Publishing Company. The oldest title is *Economic working papers bibliography (EWPB)*, which began in 1973. In 1982, it was joined by the *Management and accounting working papers bibliography (MAWPB)*. Both titles have author, series and subject indexes, and they normally only include papers received at the University within one year of the publication of the paper.

Recent developments in the University of Warwick collection, including an involvement with the Institute of Management International Databases *Plus* (IMID), have been described by Cleave (1995). Working papers and discussion papers from all sources are regularly announced in *BRTT*. The London Business School also collects working papers of its own and other academic researchers in the business field, and a *List of working papers received* is issued monthly.

Experience with working papers in the United States has been described by Soules *et al* (1992), with specific reference to activities at the Kresge Business Administration Library of the University of Michigan, where working papers are provided as an essential service because users believe access to them is crucial for their current research projects. The paper observes, however, that many libraries offer highly limited access to working papers or do not provide access at all. One reason for this state of affairs is a lack of resources to carry out the selection, acquisition and processing of the documents themselves; another is the belief that if research described in a working paper really is of value, it will eventually appear in a more formal and accessible publication.

Working papers are not, of course, the exclusive medium of the business studies specialist and economist; they are also much favoured by politicians and officials at government and international level. A good example in Great Britain was the publication in 1989 of a set of eight working papers designed to fill in key details missing from a White Paper on National Health Service reforms. Another instance of the widespread use of working papers is found at the United Nations, where several important series are made available on subscription. Among them are the *International business commerce trade and development papers*, produced by the UN Commission for Trade and Development (UNCTAD), Geneva; *Industrial development papers* produced by the UN Industrial Development Organization (UNIDO), Vienna; and the documents of the UN General Assembly and the Security Council, New York. The UNCTAD and UNIDO output includes reports and statistics which are not available anywhere else, whilst the working papers from the General Assembly and the Security Council consist of the provisional records of meetings, preliminary drafts or reports and letters, and draft resolutions.

In Europe, the European Parliament compiles many documents on business and economics, in particular the *Session documents* which embrace motions for resolutions, consultation reports, own-initiative reports, budget reports, Commission proposals, and transfers of appropriations. Details are held on the Epoque database from 1979 onwards.

For those who need to exploit collections of European Union documents, in which many items are published either in full or in part through normal publishing channels (and so cannot be truly classed as grey literature), a number of comprehensive guides are available, and some of these are discussed in Chapter 9.

Market research reports

As a form of grey literature, market research reports have a number of special features which distinguish them from other publications in the genre. Firstly, they are often expensive, or relatively so, to purchase; secondly, (like some of the products they deal with) they have a very short shelf life; and thirdly, they are regularly announced and discussed in the daily press as well as in the appropriate specialized journals.

Market research reports and market surveys are also noted to some extent in *BRTT*, in section 05D: economics and economic history. The

British Library also updates from time to time a useful compilation called *Market research: a guide to British Library collections* (Leydon and Lee, 1994).

Market research reports are excellent sources of ready worked market information, presenting analyses of data on a market sector, product or service in a structured and authoritative way which can be of immediate use to an inquirer, and in many cases does away with the need for specific desk and field surveys. Many reports on consumer information are available for the asking from the originators, and Mort (1990) has made a survey of free of charge 'grey publishing' in this area.

The British Library Science Reference and Information Service (SRIS) has a collection of over 2000 current marketing reports on a broad range of topics, built around publications issued by a core group of publishers, with other relevant reports selected and added on an individual basis. The collection is supplemented within the library by a business help desk, where staff offer assistance with the identification of market surveys, trade directories, statistics, business periodicals and trade literature.

In addition to the British Library's *Guide* mentioned above, there are several other finding aids to assist the inquirer in the search for a particular item. Firstly, there are the printed directories such as the *Guide to European market information: EC countries*, issued by the London Business School (1991). Secondly, there is a wide choice of databases and services which cover virtually every type of industry. One of the leading titles is *Findex: the worldwide directory of market research reports studies and surveys* available from Cambridge Scientific Abstracts in the United States and from Euromonitor in Great Britain. *Findex* contains lists of reports from over 700 publishers and the 18th edition (1997) provides information on over 5200 previously unlisted reports. Other important sources are the *Market studies library*, compiled by Find/SVP Inc. and providing comparative data and intelligence on the competition and product development ideas for a variety of industry segments, and Knight Ridder's *Market research locator*, which contains abstracts from over 850 US and non-US publishers, from 1985 to date, with quarterly updates. One of the many individual providers of market research reports is Management Accountancy and Computer Education (MACE), which for over ten years has been issuing a range of reports examining the UK vertical markets for information technology. A typical title is *Information systems in the UK energy industries, 1997*.

Reports from bankers and stockbrokers

Of all the grey literature documents in the business and economics area, reports based on research carried out by stockbrokers are probably the most transitory, primarily because they are normally produced for clients with the object of influencing investment decisions. Nevertheless, whilst such reports do remain current, they are often invaluable sources of information about the financial world, and are frequently quoted and commented upon. The British Library's SRIS receives reports from major stockbroking firms, including Barclays de Zoete Wedd, Alexanders Laing and Cruikshank, and Morgan Stanley. Results of stockbroker's research are increasingly becoming available online, as for example the *ICC stockbroker research database* which gives the full text of international stockbroker reports, providing detailed comparisons, assessments and projections of company performance and industry trends.

Reports from banks, on the other hand, have a more lasting value, especially those from the world's central banks, which are a key source of up-to-date financial, economic and statistical information on the countries from which they are issued. The reports comprising the collection at the Joint Library of the International Monetary Fund (IMF) and the World Bank, Washington, are available as *Annual reports of the world's central banks* (Chadwyck-Healey). The year 1984 constitutes the base date for the collection, and updates are available for following years. There is also a retrospective collection covering the period 1946 to 1983.

The types of grey literature produced by the IMF, including country specific economic information and research reports on international economic topics, have been summarized by Perry (1995). The World Bank itself publishes a number of titles, including the *World Bank's world development report 1978–1996, World data* and *World debit tables*, which are available in Great Britain through Microinfo Limited. The work of the World Bank in the field of grey literature has been reviewed by Cummings (1995). Many individual banks in all countries publish documents which can be regarded as belonging to the grey literature. Examples include *UK industry briefings* from Barclays Bank, and *Sector reports* from the Union Bank of Switzerland. Details are available on the Quest Economics Database compiled by Janet Matthews Information Services and distributed in CD-ROM format by Chadwyck-Healey.

Statistical information

Statistical information is a subject which has been extensively considered elsewhere, as for example the contribution by Mort and Clitheroe (1996) to the *Reference sources handbook*, in which they note that the vast majority of statistics in most countries is produced by national governments and their agents. Many printed series of official statistics are published in the conventional manner, and so cannot truly be regarded as part of the grey literature. Thus, taking the European Union as an example, the key publications include the *Eurostat yearbook*, (tables, graphs and maps comparing social and economic indicators over time and by country), *Basic statistics of the Community* (a comprehensive review of the European Union and its main trading partners), *Europe in figures* (an overview of the European Union) and *A Social portrait of Europe*, (a picture of society in Europe). On the other hand, a great deal of the Communities' statistical information is stored as Eurostat data in a range of different databases (Cronos, Comtext, Regio, Eurofarm, IOT and Sabine). Data are available on line, on magnetic tapes on diskettes or as print-outs.

In Great Britain, the Office for National Statistics (ONS) publishes *Government statistics: a brief guide to sources* (Newport, Gwent) which is update annually. The ONS is part of the Government Statistical services (GSS) which provides the United Kingdom with most of its official statistics. A vast amount of this information is published regularly, and more is available on request. Whether statistical data can be categorized as literature, let alone grey literature, is open to question; what is certain is that the Eurostat and GSS examples contain a mix of information available through the normal publishing channels and information which is so new or of such restricted interest that it is only available online or by application.

In the United States, a collection prepared by the Department of Commerce, Bureau of the Census is *Statistical abstracts of the United States 1995*, available from a major grey literature source, namely the National Technical Information Service, NTIS, as report PB95–965801-CDG.

On a global basis, the distinction between 'official' and 'grey' statistics has been recognized by the publication of the *World directory of non-official statistical sources* (Euromonitor, 1997), which gives details of over 3000 regularly published titles, serials and statistical data services, including university surveys and trade association membership data books.

Summary

As noted at the outset, the grey literature of business and economics mirrors to some extent that of science and technology; the one major absentee is a comprehensive announcement journal covering all types of publications in the manner, say, of *Government reports announcements and index*. The volume of material is growing apace, and in addition to the specific categories of documents mentioned above, note must be taken too of conferences, meetings and symposia, all of which give rise to papers and preprints of the types examined in Chapter 5. Reports issued by public limited companies (PLCs) and their overseas counterparts make up a further rich and varied source of information, issued primarily for the enlightenment of shareholders and employees, but always available on request from the company secretary's office. A recent trend has been the preparation of two versions of a company's report, one in the traditional comprehensive manner as prescribed by statute, and one of reduced scope and content aimed at the reader with a requirement for a more general view. Many collections of company reports are provided in major libraries, but more and more the preferred access route is via an online database such as *FT analysis reports*, which details the activities and financial results of over 5000 publicly quoted companies from the United Kingdom and a number of European countries. Increasingly, too, companies are provided access to their affairs via web sites on the Internet.

References

Cleave, G.E. (1995) Project management in grey literature: a study based on the development of the Working Papers Project at the University of Warwick. In Farace, D.J. (ed.) (1996) *Proceedings of the Second International Conference on Grey Literature, Washington 1995.* Amsterdam: TransAtlantic Publishing

Cummings, L.H. (1995) The World Bank as a source of grey literature. In Farace, D.J. (ed.) (1996) *Proceedings of the Second International Conference on Grey Literature, Washington 1995.* Amsterdam: TransAtlantic Publishing

Fletcher, J. (ed.) (1984) *Information sources in economics.* 2nd edn. London: Butterworths

Lester, R. (1996) *Information sources in finance and banking.* London: Bowker-Saur

Leydon, M. and Lee, L. (eds.) (1994) *Market research: a guide to British Library collections,* 8th edn. London: British Library

Mort, D. (1990) Free consumer market information. *Business information review,* 7, (1) 10–17

Mort, D. and Clitheroe, L. (1996) Statistical sources. In Lea, P.W. and Day, A. (eds.) *The reference sources handbook,* 4th edn. London: Library Association Publishing

Perry, B. (1995) Grey literature in the International Monetary Fund. In Farace, D.J. (ed.) (1996) *Proceedings of the Second International Conference on Grey Literature, Washington 1995.* Amsterdam: TransAtlantic Publishing

Soules, A., Lucas, J. and Pritts, S. (1992) Compromises in the management of working papers. *Library resources and technical services,* 36, (4) 478–486

Vernon, K.D.C. (ed.) (1984) *Information sources in management and business,* 2nd edn. London: Butterworths

CHAPTER NINE

The European Community

Introduction

The European Community (EC) has had a direct interest in grey literature ever since the initiative taken at the York Conference of 1978 to set up the SIGLE database (see Chapter 1). Since that time, the EC involvement in the area has grown considerably, and quite apart from the broad brush developments described in this chapter, there are other references scattered throughout the book to the contributions made by various EC organizations. See, in particular, *Euro abstracts*, which announces reports arising from Research and Technological Development programmes, and is described in Chapter 12 dealing with science and technology.

In looking at the documentary output of the EC, it is necessary to go back to the very beginning. The origins of the European Community lay in the desire following the Second World War to replace the European system of competing nation states with a new economic union. It was partially out of a wish to heal traditional German-French enmities that in May 1951, Robert Schuman, the French foreign minister of the time, proposed that France and West Germany (as it then was) pool their coal and steel industries under an independent authority. Belgium, Luxembourg, the Netherlands and Italy quickly joined in, and the first Community, the European Coal and Steel Community (ECSC) was established in 1952, following the enabling Treaty of Paris, 1951. Further moves towards European economic integration ensued with the Treaty of Rome, 1957, which established the main European Community. At the same time, a second treaty founding a third Community, the European Atomic Energy Community (EURATOM) was signed, pledging the signatories to cooperate in research into nuclear science and in particular the peaceful uses of atomic energy. The Treaty of Rome was intended to create a customs union to remove all obstacles to the free movement of capital, goods, people and services between member states. The Treaty also established the Community's institutional structure, namely the Commission, the Council of Ministers, the Economic and Social Committee, the European Investment Bank, the Parliament and the Court of Justice.

In June 1970, the original six member states invited Great Britain, Denmark, Ireland and Norway to open negotiations on their applications to join the EC. All four countries signed a Treaty of Accession in Brussels in 1972, although Norway subsequently withdrew following the negative outcome of a referendum. The enlarged Community of nine came into being in 1973. Greece was admitted in 1981, and Portugal and Spain joined in 1986. The Treaty of European Unions (the Maastricht Treaty) was signed in 1992 and formally established the European Union (EU). The number of member states contributing to the EU budget has now grown to fifteen, and the full list reads: Austria, Belgium, Denmark, Finland, France, Germany, Greece, Ireland, Italy, Luxembourg, the Netherlands, Portugal, Spain, Sweden and the United Kingdom. Applications from a number of other states are under consideration.

The growth in the size and activities of the European Union has been matched by an enormous growth in the output of EU documents, a growth which itself has been compounded by the need to publish material in various language versions, each identified by a language prefix, namely: ES – Spanish; DA – Danish; DE – German; GA – Gaelic; GR – Greek; EN – English; FR – French; IT – Italian; NL – Dutch; PT – Portuguese; FI – Finnish; and SV – Swedish.

Guides to EU publications and activities

Just as over the years the flow of publications originating within the European Union has grown from a manageable trickle to a mighty flood, so in parallel with this increasing output, the externally produced guides to what these publications contain has also grown from a handful of explanatory texts to a wide range of specialized surveys, summaries and directories. Typical of such titles are *The European Community* (Jones and Budd, 1994), which explains how to set up a good basic library of European official material at a low cost, the *Penguin companion to the European Union* (Bainbridge and Teasdale, 1995), which indicates secondary sources for topics frequently identified by shorthand designations, such as the 'Cockfield paper', the 'Kangaroo Group' and the 'Schengen Agreement', and *European Union information* (Ramsay, 1997), a brief guide to basic printed and electronic sources. A comprehensive guide to publications and other topics, which first appeared in 1989, is the *Directory of EU information sources (The Red Book)* (Euroconfidentiel, 1997), which provides facts and figures on all the EU institutions, with details of the key personnel, databases, publications, information networks and libraries specific to each institution. Details are also given of information and documentation centres.

In addition to the many such publications compiled by and for users of EU documents, there have been significant developments in organizations specializing in the information resources of the EU, and most member states now have a group of documentation professionals monitoring activities and exchanging know-how. For example, The European Information Association (EIA) concentrates on matters relating to librarians and information officers, whilst the European Association of Information Services (Eusidic) has members representing publishers, online service vendors, government departments, software houses, learned societies, database producers, university libraries, telecommunications companies, lawyers, and information brokers. Both associations hold annual conferences, with the activities of the EIA the subject of a regular column in the Library Association's quarterly journal *Refer*, and those of Eusidic reported in the organization's own newsletter *Newsidic*, issued bimonthly.

Thus, on the one hand, there is no shortage of externally published information about the European Union nor, on the other hand, is there any lack of organizations concerned with interpreting and exploiting the EU's publications. And, as will be seen, a great deal of help is available from the European Union itself. In fact, the twin themes of literature published by EU institutions and literature written about the EU are inseparable and have been examined by a number of workers, as for example Schlogl (1990), who gives the Austrian viewpoint.

European Union publications

The current awareness key to the European Union's own output of publications is *EUR-OP NEWS*, a reader-friendly quarterly newsletter from the European Communities' Publications Office. The range of topics covered is considerable, as the following list of subject headings regularly in use clearly shows:

1	Agriculture	16	Fisheries
2	Audio-visual	17	Foreign trade
3	Citizens' Europe	18	Information policy
4	Competition policy	19	Information society
5	Court of Justice	20	Internal market
6	Culture	21	Monetary union
7	Eastern Europe	22	News from EUR-OP

8 Economy	23 News from the institutions
9 Education	24 Regional and development policy
10 Employment	25 Research
11 Energy	26 Small and medium enterprises
12 Environment	27 Social policy
13 Equal opportunities	28 Telecommunications
14 External market	29 Transport
15 External policy	30 Women

The correct and precise identification of the items making up such a diverse publications programme is essential, and the various document identifiers used by the EU follow the traditional grey literature practice of a mixture of letters and numbers calling for careful transcription to ensure that any items requested are properly provided. Some typical examples of the identification system for EU publications are given below:

Beer delivery agreement in the new member states (CV-01–96–074-DE-C);
Evaluation of the EU aeronautics programme (1990–1994) (CG-NA-16–984-EN-C);
Working conditions in hospitals in the EU (SY-93–95–499-EN-C).

Many of the publications listed in *EUR-OP NEWS* carry the abbreviations *OJ* and *COM* which stand respectively for the *Official journal of the European Communities* and *Commission documents*. Thus *Washer-dryers:* New Directive 96/60/EC implementing Directive 92/75/EEC on energy labelling is identified as *OJ L 266 1996*, whilst a Commission Green Paper on noise reduction is listed as *COM (96) 569 fin. Commission documents* are, in general, proposals for legal acts not yet published in the *Official journal*. The *OJ* also has a *Supplement* (designated *OJ/S*) which features on average 500 public tender notices every day, and more than 140 000 each year. The *Supplement* is available on subscription as a CD-ROM (FX-AS-97–000–1F -Z). Most EU publications are available for sale, with cover prices given in ECUs, via national sales agents in the member states. Some publications listed in *EUR-OP NEWS* are available free of charge, and can be obtained via the information offices of the European Commission and the European Parliament established in the member states. *EUR-OP* also offers a range of online services, including:

ABEL (Amtsblatt elektronisch), a document delivery system for EU directives;
CELEX (Communitas Europaea lex), for matters relating to EU law;
INFO 92, for details of the Single Market and the Social Charter;
SCAD (Systeme Communautaire d'Acces a la Documentation), for information on EU policy.

A complete list of databases is contained in the *European Union database directory* (JX-98-96-639-EN-C).

Ready access to EU documentation is available through the long established network of European Documentation Centres (EDCs), over 300 of which have been set up in universities throughout Europe and beyond. One of the most important EDCs is that hosted by the European Institute of Public Administration (EIPA) in Maastricht, which offers access to all *EUR-OP* publications and documents. Among the EIPA's most recent publications are a *Guide to official information of the EU* and *The EU's common foreign and security policy: the challenges of the future*. For a full list of EDC's and other EU information services, the reader is referred to *Europe info* (CC-88-95-767-3A-C).

A more informal level of access to EU documents is provided by the Public Information Relays (PIRs), based in selected public libraries throughout the Community, which receive and display some of the information and materials (particularly pamphlets and maps) published by EU institutions and other relevant bodies. Like many other large publishing organizations, the EU is now making full use of the Internet, and in 1997 issued *EUR-OP*'s first Internet and electronic products guide, *Europe goes electronic*. The compilation contains comprehensive details of more than 80 Internet addresses and electronic products, both off-line and on-line; the guide is available from any of the *EUR-OP* gateways.

Summary

As expressed by Jacques Santer, President of the European Community, 'openness, communication and information' are key words in the process of bringing the European Union closer to its citizens. *EUR-OP* serves as the official publisher for the Union's institutions, bodies and agencies, and lies at the heart of the Union's information production and dissemination activities. *EUR-OP* has, as a multilingual publishing body, the twin responsibilities of focusing on the cultural and linguistic heritage of each of the member states whilst contributing at the same time to the development of a new European consciousness. Nevertheless, despite the substantial efforts to publicize and make available the EU's information resources, and to render them easier to access, many professionals who spend their working lives identifying and distributing EU information to their clients insist that to succeed it is necessary to know the system (Lyon, 1996). This is especially true where advance information is important, such as that dealing with proposed legislation. Perhaps in response to just such sentiments the European Union is now running a four-year programme (1996–1999) called *INFO 2000*, with support for a range of activities which include exploiting the wealth of European content (words and pictures, sounds and images, facts and ideas), producing attractive products and services for electronic information consumers, cultivating new creative and commercial skills, and creating new jobs and information markets. *INFO 2000* is administered by Directorate General DG XIII: Telecommunications, Information Market and Exploitation of Research. The slogan of the programme is 'information is everybody's business' – let us hope it succeeds.

References

Bainbridge, T. and Teasdale A. (1995) *The Penguin companion to the European Union.* London: Penguin Books

Euroconfidentiel (1997) *Directory of EU information sources*, 8th edn. (*The Red Book*). Brussels: Euroconfidentiel S.A

Jones, A. and Budd, S. (1994) *The European Community: a guide through the maze*, 5th edn. London: Kogan Page

Lyon, J. (1996) Searching for your needle in the EC haystack. *Information world review*, December , 37–38

Ramsay, A. (1997) *European Union information*, 2nd edn. London: Association of Assistant Librarians

Schlogl, J. (1990) Acquisition of EU literature and grey literature. *Mitteilungen der VOB*, **43**, (4) 91–98

CHAPTER TEN

Education

Introduction

All the perspectives of study in grey literature in education employ their own particular terminology to make a point. The same terms have different meanings much more often than in the language of science, and a tendency for the idiosyncratic use of conceptual language reflects the long tradition of individual research in education and allied subjects. In education, too, the tangle of terminology is made denser still by differences between national systems of education and changing schools of thought. Furthermore, educational research, whether of the extended study or snapshot survey kind, becomes obsolete through social changes beyond its control, rather than through the momentum of its own progress, as is the case in science.

Important studies in education, as was noted in the first edition of this work (Davies and Gwilliam, 1975), are certainly published much less frequently in technical report series than are studies in conventional scientific fields. Even when they are issued, they do not exhibit the typical features of consistent serial organization and laboratory production, instanced even in work on education's foundation subjects such as applied psychology, psychometrics and other areas

The conventional scientist probably has greater incentive to disseminate his findings and greater opportunity to do so formally. However, the importance of informal contacts for the economical but adequate exchange of information among educational workers should not obscure the fact that similar kinds of communication are typical of all scientific fields, one term for which is the 'invisible college'.

The place of the unpublished document in educational studies (traditionally somewhat lacking in formal organization) is conditioned by the investigator's concern to control, if not actually restrict, the personal exchange of information in ways which seem to be the most productive for them, and also the most conducive to an understanding and informed use of the results reported.

There is undoubtedly a large proportion of reports in education with a small technical content, many dealing with current development, policy and administrative matters, and some attempting to inform non-specialists about teaching and educational research and their mutual relevance. Many reports and other unpublished documents have an essentially local value; many quickly become obsolete. Yet collectively, such items become more valuable to historians as social source documents rather than as scientific and technical expositions, if only because education comprehends and reflects so many aspects of social evolution and social need, including indeed the changing face of science and technology.

As a conclusion to this brief look at the background to reports information and its treatment in the world of education, reference must be made to the book by Dibden and Tomlinson (1981) entitled *Information sources in education and work*, which adopts a somewhat different approach from other titles in the *Information sources in . . .* series since it confines its attention to publications and sources likely to help anyone interested either in starting a career or involved with assisting those who have to make a career choice, and so is not aimed at the information specialist.

Document sources

The origins of reports and other items of grey literature in the field of education can be traced in official publications at national and local government level from the mid-19th century onwards. The appearance of such reports coincides with the period during which educational studies became autonomous subjects at certain universities. As with many other subject areas, much of the most useful grey literature material receives more formal publication at a later stage in its life, but of course a great deal does not. Furthermore, although there is a substantial continuity in original sources, many documents derive *ad hoc* from particular projects, programmes and individual efforts. This trend shows, at least in Great Britain, every sign of continuing, for the European Policy Forum (Stewart, 1992) notes: 'A new magistracy is being created in the sense that a non-elected elite is assuming responsibility for a large part of local governance. They are found on the boards of health authorities and hospital trusts, Training and Enterprise Councils, the boards of governors of grant-maintained schools, the governing bodies of colleges of further education, and Housing Action Trusts'.

An examination of the educational scene reveals that the following types of organizations issue a whole range of grey literature:

(1) national public bodies, such as the Department for Education and Employment, the Economic and Social Research Council, and the National Curriculum Council;
(2) local public bodies, such as local education authorities;
(3) foundations, including the great charitable organizations such as Nuffield, Ford, Rowntree, Leverhulme, Calouste Gulbenkian and Van Leer;
(4) universities and colleges, especially those establishments with departments of education, psychology and/or sociology, and particularly those institutions possessing separate schools of education;
(5) regional and international organizations, as for example the United Nations Educational Scientific and Cultural Organization (UNESCO);
(6) associations and professional organizations, of which there are very many – for instance the *Education Authorities' Directory and Annual* (School Government Publishing, January each year), has a section called 'Educational associations, societies and other organizations concerned with education', with a list running to over 800 different bodies, many of which issue publications of one sort or another. The names range from the expected (the Association for Science Education, the National Association for Gifted Children, and the School Library Association) to the not-so-expected (the Association of British Insurers, the National Dairymen's Association and the Royal Philatelic Society).

ERIC Announcement services

The grey literature relating to education, mainly but not exclusively that originating in the United States, is announced in the monthly *Resources in education (RIE)*, which until 1975 was known by the title *Research in education*. *RIE* is sponsored by the Educational Resources Information Center (ERIC) of the Office of Educational Research and Improvement (OERI), part of the US Department of Education, Washington. ERIC constitutes a nationwide information network for acquiring, selecting, abstracting, indexing, storing, retrieving and disseminating education-related documents considered timely and significant. The network embodies a coordinating staff in Washington and sixteen Clearinghouses located at universities or with professional organizations across the United States. The Clearinghouses, each responsible for a specific educational area, are an integral part of the ERIC system, and their names are:

(1) Adult Career and Vocational Education (CE);
(2) Counselling and Personnel Services (CG);
(3) Educational Management (EA);
(4) Elementary and Early Childhood Education (PS);
(5) Handicapped and Gifted Children (EC);
(6) Higher Education (HE);
(7) Information Resources (IR);
(8) Junior Colleges (JC);
(9) Languages and Linguistics (FL);
(10) Reading and Communication Skills (CS);
(11) Rural Education and Small Schools (RO);
(12) Science Mathematics and Environmental Education (SE);
(13) Social Studies / Social Science Education (SO);
(14) Teacher Education (SP);
(15) Tests Management and Evaluation (TE);
(16) Urban Education (UD).

Resources in education consists of comprehensive document descriptions (termed document resumes) and indexes, The résumés provide detailed accounts of each document, and are numbered sequentially by an accession number beginning with the prefix ED (which stands for ERIC Document). The indexes provide access by subject, personal author, institution and publication types. The last named index is of special interest because it characterizes documents by their format as distinct from their subject matter, and so presents a comprehensive overview of the components of grey literature in the field of education. The publication type categories are summarized by code and category as follows:

010 Books, collected works (including conference proceedings);
030 Creative works, dissertations/theses;
050 Guides;
060 Historical materials;
070 Information analyses;
080 Journal articles;
090 Legal/legislative/regulatory materials;
100 Audio-visual/non-print materials;
110 Statistical data;
120 Viewpoints (opinion papers, position papers, essays, etc.);
130 Reference materials;
140 Reports;
150 Speeches, conference papers;
160 Tests, evaluation instruments;
170 Translations.

The documents cited in *RIE* are, except where otherwise noted, available from the ERIC Document Reproduction Service (EDRS) in both microfiche and paper copy or, in some cases, microfiche only. In the United Kingdom, ERIC documents can be obtained from the British Library Document Supply Centre. Each ED entry in *RIE* constitutes a carefully formatted record, including author, title, the name of the organization where the document originated, the date published and the contract or grant number. Also given is the name of the sponsoring agency, the report number, descriptive notes, a list of descriptors, and an informative abstract with the abstractor's initials. The subject coverage of *RIE* is wide and ranges from academic achievement to youth employment, and from equal opportunities to school desegregation. The institutions listed in the index include the Department of Education itself, foundations, universities, associations, and certain newspapers.

As with many other disciplines, proper acquaintance with the literature of marginal fields is necessary to exploit the core subject, and in this respect a large information system such as ERIC can achieve adequate coverage not only of statistics, psychology, sociology and economics as appropriate, but also relevant studies in vocational guidance and training, employment, cybernetics and computers, plus librarianship and information science.

The bibliographical database which embraces ERIC activities consists of two files, namely *Resources in education (RIE)* as noted above, and *Current index to journals in education (CIJE)*, which complements *RIE* and treats published journal literature from over 775 periodicals. The database records go back to 1966 and are accessible via several hosts and also in CD-ROM format from Silver-Platter. Records are approaching a total of one million items, and over 31 000 new items are added each year. The database also includes the full texts of *ERIC* digest records, one or two page documents aimed specifically at teachers, administrators and other practitioners to give an overview of a given topic, with references to follow-up information.

A number of guides to ERIC are available: see, for example *Steps in using ERIC* (ED 288 528) and *A parent's guide to the ERIC database* (ED 340541), whilst aids to searching procedures include *Indexing and retrieval in ERIC* (ED 279 346) and the *Thesaurus of ERIC descriptors* (Oryx Press, 1995).

Most of the documents described in *RIE* are channelled into the system via one of the Clearinghouses, but each issue of *RIE* also carries an open invitation to document producers to submit details of 'unpublished or 'fugitive' material not usually available through conventional library channels'. Users are also encouraged to contact ERIC direct, when the experience can be rewarding. Indeed, one user (Latta, 1992) describes searching the ERIC database as an 'educational adventure'.

Other announcement services

The sheer size and resources of ERIC dwarf efforts by other agencies to tackle the question of grey literature in education, but the initiative taken by the British Library Document Supply Centre in seeking to coordinate items originating from British sources is growing in significance. The announcement journal *British reports translations and theses (BRTT)* lists items relating to education and allied topics under some of the headings which form part of section 05: Humanities, psychology and social sciences, notably the following:

05B Documentation, information science and librarianship;
05K Linguistics;
05P Education and training;
05Q Psychology;
05R Sociology.

Topics covered range from *Changing primary schools* (92–09–05P-029) to *Art in the National Curriculum* (92–09–05P-027), and from *Standards of arithmetic – how to correct the decline* (96–11–05P-072) to *Professional development to meet special educational needs* (96–11–05P-089). The announcement services familiar to workers in science and technology also carry items relating to educational matters. *Government reports announcements and index (GRA&I)* has a place for education under the category 'Behaviour and society'. It also reports documents with an educational content under other headings, notably 'Library and information sciences'. A typical entry in *GRA&I* is *Virtual reality and education* (AD-P006 942/7/GAR). *Scientific and technical aerospace reports (STAR)* places educational topics under category 80: Social sciences (general), but entries here are generally sparse.

The general picture which emerges is that apart from *BRTT*, most grey literature announcement services feature very little on education, and seem content to leave the garnering of publications in this area to the ERIC system. This approach is emphasized by a further ERIC activity, for in addition to collecting and collating the literature of education for announcement in *RIE* and *CIJE*, the ERIC Clearinghouses also analyse and synthesise the literature into a number of different formats which are designed to compress the vast amounts of information available into manageable portions, and so to meet the different needs of ERIC users. These formats include, apart from the *Digests* mentioned above, research reviews, state-of-the-art studies, interpretive studies on topics of high current interest, research briefs, annotated bibliographies, and other compilations. Such documents extend from a short note on *Delivering special education* (ED 340 154) to a 66-page work, the *Distance education handbook* (ED 340 547).

Despite the undoubted excellence of ERIC, some educational specialists in the United Kingdom feel that the service does not offer an altogether adequate coverage of British research and resources. The reader can, however, turn to a compilation prepared by the National Foundation for Educational Research in England and Wales (NFER), namely the *Register of educational research in the United Kingdom*. The *Register*, which dates from 1973, is also available via EUDISED, discussed below. In addition to the preparation of the *Register*, NFER is an important source of research reports in its own right, as for example *What teachers in training are taught about reading* (1992) and *What DO students thinks about school?* (1993).

A short-lived development in the availability of British sources was the National Educational Resources Information Service (NERIS), which provided resources supporting the National Curriculum and the curriculum requirements in Scotland, Northern Ireland and Wales. NERIS, which was set up in 1986 and launched as a CD-ROM service in 1989, was also available online, and drew its information from more than 1250 different, validated sources, including books and grey literature items such as reports, notes and worksheets. For example, reports and papers listed in the British Library Research and Development Department's catalogue *Information skills in education* (1990) appeared in the NERIS files. Such, in fact, was the variety of inputs to NERIS that a subject inquiry, for example on the topic of rain forests, would result in the identification of dozens of items ranging from books, videos, and TV broadcasts to journal articles and resource packs. Despite a promising start and the support of over 2000 subscribers, the Department for Education announced in 1993 the withdrawal of support for NERIS because of a failure to meet a series of financial targets.

As was mentioned earlier in this chapter, a great deal of information relating to education is published by governments and international organizations; much of it does belong, of course, to the grey literature because it is announced and made available through, for example, The Stationery Office (TSO) (formerly Her Majesty's Stationery Office, HMSO) or the United States Government Printing Office (USGPO). Indeed, TSO publishes sufficient material on education to warrant a special *Education catalogue*. Similarly, the publications on educational and cultural policy issued by the United Nations and the European Community can be identified respectively in *Unesco bulletin* and *EUR-OP NEWS*. If, however, publication takes place outside the main channels of dissemination, specialist services need to be checked, and in the case of Great Britain, a major source is Chadwyck-Healey's *Catalogue of British official publications not published by HMSO* (now called *UKOP*, see Chapter 1), which lists titles such as *Public libraries and adult independent learners* (Council for Educational Technology) and *Career breaks for women engineers and technicians* (Engineering Industry Training Board). On broader issues, the monthly *International labour documentation* published by the International Labour Office, carries details of monographs as well as journal literature, and includes in its subject coverage vocational training and social development.

EUDISED

An important service in Europe which is concerned with education is the European Documentation and Information System for Education (EUDISED), the formative stages of which were described by Davies and Gwilliam (1975). EUDISED is a project of the Council for Cultural Cooperation of the Council of Europe for a decentralised computer-based education documentation and information network. At the beginning, the principal concern was the recording of on-going research, and in particular, coverage of ERIC report listings. European report sources were to be covered economically by project entries in a standard format, organized by a special multilingual thesaurus giving details of the literature associated with the projects. Later, it was proposed to enlarge EUDISED beyond the coverage of grey literature by adding details of articles from books and periodicals dealing with education matters. It is now possible to access EUDISED as an online database through the host ESA/IRS. The time span is 1975 to date, the references cover compulsory, higher and adult education; vocational training; principles, systems and administration of education; psychology; sociology, philosophy and economics of education; teaching methods and aids; personality development; and educational information sources.

An appraisal paper by Davies (1980) noted that problems existed, namely 'difficulties in getting national cooperation, the necessity of unravelling relationships with other international organizations in a concrete and constructive way with a minimum of overlaps and unnecessary repetitions and competitive ventures, the paucity of staff, and the inability to attract appropriate funds for essential activities'. When first conceived, the approach to EUDISED was seen as similar to but not identical with the ERIC system, and the heart of the scheme was a measure to tackle the not inconsiderable language problems. The solution was the *Multilingual thesaurus for information processing in the field of education*, the first edition of which was published under the aegis of the Council of Europe in English, French and German versions in 1973, in a Spanish version in 1975 and in a Dutch version in 1977. A second edition appeared in 1984, and the third edition came out in 1991 as the *European education thesaurus* from the Office for Official Publications of the European Communities.

Representatives from around 30 countries are currently involved in developing and refining the EUDISED system, which is also available in a printed version as *EUDISED R&D bulletin*. Each quarterly issue carries about 250 entries, with the remainder of the 25000 or so entries added to EUDISED each year being processed directly into the database. A further development has been the inclusion of the whole of *United Kingdom register of educational research*, although as noted above, this important NFER information source will continue to appear independently as a hard copy version. A general account of EUDISED, and in particular the plans for the future, has been provided by Cosgrove (1992). As with many other database providers, the EUDISED system is now accessible on the Internet at the Biblioteca di Documentazione Pedagogica (BDP) site in Florence, Italy.

Another educational thesaurus is the *British education thesaurus (BET)*, which contains 8200 terms in current academic and professional usage, and which is used in conjunction with the *British education theses index (BETI)*, a reference source coordinated at Leeds University and containing records of theses on all aspects of education accepted for higher degrees by universities in Great Britain and Ireland. Also available from Leeds is the *British education index (BEI)* which provides regular information about the contents of over 350 British education and training journals, with a growing coverage of reports, conference papers and other monographic literature. *BEI* is available in several forms, including a quarterly periodical with author and subject indexes; British Education and Training Listing Services (*BETlists*); online via Dialog File 21; and as the British component of the Knight-Ridder International ERIC CD-ROM, which contains the Australian and Canadian education *Indexes* as well. In Germany, the Erlangen special collection in the field of educational research pays special attention to grey literature unlisted in national bibliographies, and has been described by Theuerkauf (1996).

Public involvement

Education, unlike many of the other topics discussed when reviewing grey literature, is something every reader has experienced at first hand. It is no accident, therefore, that much of the unconventional, outside the book trade type of literature in education originates at or concerns the grass roots level – because we have all been through various stages of education we are likely to have strong views as parents, taxpayers and, increasingly, as school governors. The daily press finds sufficient interest among its readers to warrant regular features, supplements and special issues devoted solely to education. There is an ever growing fund of information, comment and opinion, and some of the publications which chronicle it will eventually be recorded in one of the major databases, abstracted and indexed for subsequent retrieval; a great many, however, will simply fall into oblivion, ignored because they seem transient or trivial. Just what is collected and recorded will depend on a combination of chance and recognition. No one can formulate a strict policy on what to look for, but librarians and documentalists can influence the scope and coverage of grey literature by encouraging issuing bodies to send copies for inclusion in the appropriate national monitoring service.

The variety of titles which continue to appear emphasizes the wide range of parties with an interest in education – for example the Chartered Institute of Public Finance and Accountancy advocates the evaluation of teachers in its report *Performance indicators for schools*, whilst the Audit Commission has published a list of performance indicators relating to the *Citizen's Charter* with questions such as: 'How many children receive nursery education? How full are secondary schools? Can children eat hot meals at school?'

The Equal Opportunities Commission has discovered and written about sex discrimination in West Glamorgan schools; the International Freedom Foundation proposes encouraging children to join voluntary defence training corps in its report *Education for defeat;* the Institute of Economic Affairs wants the government to introduce education tax credits; the *Survey of comics and magazines for children and young people* does not recommend Enid Blyton's *Famous five adventure magazine* because of alleged sexual stereotyping; and the Centre for Policy Studies alleges there is something wrong in the classroom with the pamphlet *Teachers mistaught: training in theories or education in subjects?* The list can be extended indefinitely, and serves merely to remind searchers that no surveys or studies of publications on education and related matters should be considered complete unless the net has been cast as widely as possible and taken in grey literature sources. Regrettably, not all reports summarized in the general press are identified by a title or other bibliographical details, so making the documents in question somewhat difficult to cite correctly. See, for example, the lengthy appraisals in *The Times* (O'Leary, 1997) and the *Daily Telegraph* (Clare, 1997) of a study on the General National Vocational Qualification (GNVQ) conducted at the London University Institute of Education.

References

British Library (1990) *Information skills in education*. London; BL Research and Development Department

Clare, J. (1997) Vocational studies 'fail to meet needs for job market'. *The Daily Telegraph*, 6 June 1997

Cosgrove, A. (1992) EUDISED: European Documentation and Information System for Education. *Education libraries journal*, **35**, (1) Spring, 1–9

Davies, J. and Gwilliam, A.B. (1975) Technical reports in education. In Auger, C.P. (ed.) *Use of reports literature* London: Butterworths (Chapter 10)

Davies, J. (1980) EUDISED – image and reality: a crisis of identity. *Education libraries bulletin*, **23**, (3) 1–15

Dibden, K. and Tomlinson, J. (1981) *Information sources in education and work.* London: Butterworths

Latta, A. (1992) Educational adventure: searching the ERIC database. *DLA bulletin*, **12**, (2) Fall 6–9

O'Leary, J. (1997) Vocations and doubts. *The Times*, 6 June 1997

Stewart, J. (1992) *Accountability to the public.* London: European Policy Forum

Theuerkauf, J. (1996) The Erlangen special collection. *Bibliotheksforum Bayern*, **24**, (1) 27–43

CHAPTER ELEVEN

Energy

Introduction

Public complacency about energy and the supply of fuel received a sharp jolt as a result of the 1973–74 oil crisis, when the western world realized that the days of 'cheap' fuel were over. Governments around the world turned their attention to energy technology and how it could be used to conserve and better utilize existing resources. Attention was also turned to alternative renewable sources such as wave power and solar energy, and these activities resulted in the establishment of national and international agencies, research and development programmes, and a greater concern for the state of fossil fuel reserves. The United Kingdom was in a fortunate position in that it was able to establish itself as having the largest energy resource of any country in the European Community, and since 1980 has been self-sufficient in energy in net terms. Paradoxically, the Department of Energy, which once had responsibility within the government for the development of policies in relation to all forms of energy, no longer exists, and its remit now forms part of the Department of Trade and Industry. This is in marked contrast to the situation in the United States, where the Department of Energy (DoE) continues to provide a clear focus on energy matters.

As interest and concern relating to the energy field have continued to grow, so has the associated technical literature, and recent developments have been described by McMullan (1996) in his chapter *Energy technology* in *Information sources in engineering*, and by Harris (1996) in his chapter *Nuclear engineering* in the same work. On a much broader scale, the picture has been painted by Anthony (1988) in *Information sources in energy*, and by Eagle and Deschamps (1997) in *Information sources in environmental protection*. Berkovitch (1996) has produced a guide which covers each of the main energy sources, including coal, petroleum, gas, nuclear power, and renewables in an overview of the literature which takes note of some of the problems raised at the 1992 Rio de Janeiro conference on environmental issues. Reports literature, and more recently grey literature, has always featured prominently in the energy scene, in the main because of its origins in the nuclear industry. It is necessary to look back at these origins in order to understand the present day arrangements for handling information and publications dealing with energy in all its ramifications.

Nuclear energy

Atomic energy programmes embody nuclear science and nuclear technology, which are generally defined as follows: nuclear science is the study of the production, properties and phenomena of atomic nuclei, sub-atomic particles, gamma rays and nuclear X-rays; nuclear technology is the application of nuclear sciences to other sciences and engineering, and conversely the application of other sciences and engineering to the problems of nuclear science. The grey literature discussed here is mainly unclassified reports literature, that is reports and other documents without any security, commercial or other restrictions on dissemination and availability. Reports, as has been noted elsewhere in this volume, are generally identifiable by their alphanumeric codes, but as a rule this may not be sufficient to distinguish a report from a conference paper, from a patent, from a translation, or in some cases from a journal article. For this reason, the boundaries of reports literature, including that in the atomic energy field, will vary according to individual interpretation.

It is true to say that nuclear science and its enormous output of literature are unique in at least some respects among the various scientific disciplines. Apart from some fundamental aspects of nuclear physics which had been subject to steady scrutiny for a large number of years, nuclear science was born virtually overnight early in 1939, with the publication of letters to the editor in *Physical review*. During the summer of that year, the implications of nuclear fission were spelt out by Albert Einstein to President Roosevelt, and the reasons why atomic research during the ensuing war years had to be kept so closely guarded a secret are all too well understood today.

The initial research and development work was carried out mainly by scientists from the United States and the United Kingdom and its dominions, and the activity was known as the Manhattan project (Groves, 1962); the project was directed towards the production of an atomic bomb. Once the potential of the weapon was understood (and, more importantly, demonstrated), it was not surprising that nations with a scientific capability worked to continue safeguarding their newly acquired nuclear knowledge and to make documentation about it available only in a closely controlled manner. The method chosen for the dissemination of this information was the technical report. Gradually, controls were transferred from military to civilian agencies, and the US Atomic Energy Act of 1946 established the United States Atomic Energy Commission. The Commission's Technical Information Service was given the task of the 'dissemination of scientific and technical information as called for by the Act'.

Scientists from the United Kingdom, Canada, Australia, France and other countries, having returned home from their wartime work on the Manhattan project, became leaders and organizers of atomic energy research in their own countries. By 1954, less than ten years after Hiroshima, it was realized that the world could face extinction by a nuclear holocaust, and attention was turned to the exploitation of atomic energy for peaceful purposes. Under the auspices of the United Nations, a number of conferences on the peaceful use of atomic energy duly took place, and as a result of decisions taken at the conference held in 1956, the International Atomic Energy Agency (IAEA) was established in 1957 with its headquarters in Vienna. Today, there are over one hundred members participating in the work of the Agency, the aims of which are to accelerate and enlarge the contribution of atomic energy to peace, health and prosperity, and to ensure that any assistance provided by it or under its supervision is not used for military purposes. In the field of documentation, these aims are carried out through the International Nuclear Information System (INIS), described later.

United States Atomic Energy Commission

The development and growth of the United States Atomic Energy Commission (USAEC) has been described by a number of authorities, notably Hewlett and co-workers (1962), (1969), Smyth (1945), Fermi (1954) and Crewe and Katz (1969). The Commission's story came to an end in 1974 when the body was abolished and its functions were transferred to the US Nuclear Regulatory Commission (NRC) (Murray, 1988).

The reports literature concerned with atomic energy has its beginnings in 1942, when from that date the University of Chicago Metallurgical Laboratory provided a dissemination and central indexing system for the documents arising from the Manhattan project. During the war, the distribution of that literature was very closely guarded indeed. Then, in 1946, the USAEC was established, and the US Army and the Office of Scientific Research and Development (OSRD) relinquished control of the Manhattan Project, leaving the task of declassification and release of documents accumulated during the wartime research to the newly formed Commission. The Manhattan District Declassification Center was set up, and the first reports for public use were released (the *MDDC* and *AECD series*).

From its inception, the USAEC's research and development efforts were conducted almost entirely under contract with USAEC-owned but privately operated laboratories, with universities and non-profit institutions, or with commercial enterprises. The contractors were required to report periodically on the results or progress of their work. In basic research much of this reporting took the form of articles in scholarly journals, but the results of sponsored research were more frequently given reports submitted directly to the USAEC. Most of the reports, as well as the articles, were announced and abstracted in *Nuclear science abstracts*, published semi-monthly. The staff of the USAEC's Technical Information center was charged with the task of carrying out Section 141 of the Atomic Energy Act of 1954, which stated: 'The dissemination of scientific and technical information relating to atomic energy should be permitted and encouraged so as to provide the free interchange of ideas and criticism which is essential to progress and public understanding, and to enlarge the fund of technical information'.

Section 141 also opened wide the door for international cooperation. In the mid-1950s, the USAEC designated over one hundred centres within and outside the United States as depository libraries. Until 1963, the depository institutions automatically received either hard copy or microcard versions of USAEC sponsored R&D reports announced in *Nuclear science abstracts*. In 1963, the dissemination of reports was changed to the more convenient microfiche. For various reasons, by the early 1970s the USAEC depository system was gradually phased out. Instead, the National Technical Information Service in the United States, the International Atomic Energy Agency in Vienna, and the British Library in Great Britain became the main sources of USAEC and other reports, although some of the depository institutions continued to maintain their collections by purchase.

Energy research abstracts

The most important continuing bibliography covering the entire energy field is *Energy research abstracts (ERA)*. *ERA* provides abstracting and indexing coverage of all scientific and technical reports and patent applications originated by the US Department of Energy, its laboratories, energy centres and contractors, as well as theses and conference papers and proceedings issued by these organizations in report format. The definition now includes audiovisual material, computer media (magnetic tapes, diskettes, and so on) and engineering drawings. *ERA* also covers other energy information prepared in report form by federal and government organizations, foreign governments, and domestic and foreign universities and research organizations, provided that the full text of the document has been received by the DoE's Office of Scientific and Technical Information (OSTI).

Foreign report information for *ERA* is obtained via the *Energy database* set up by the Energy Technology Data Exchange (EDTE). The Exchange is part of the International Energy Agency (IEA), and the database carries over one million records covering the period 1987 to the present. The IEA itself was founded in 1974 as the forum for coordinating the energy policies of over twenty industrialized countries, and its publications are available through the sales agencies of the Organization for Economic Cooperation and Development (OECD). The *Energy database* contains the *INIS database* (see below) as a subset and can be accessed using an online thesaurus of controlled terms from 1974 onwards via STN, and on CD-ROM from 1987 onwards via SilverPlatter. The stated purpose of *Energy research abstracts* is to announce documents produced or obtained by the DoE that are not so readily available as journal articles, books or patents. In 1991, *ERA* stopped entirely the coverage of non-report literature.

The scope of *ERA* encompasses DoE's research, development, demonstration and technology programmes resulting from its broad charter for energy sources, supplies, safety, environmental impacts and regulation. The citations presented in *ERA* are available for searching in the *Energy science and technology database*, prepared by the DoE as a multidisciplinary file containing approximately 3.25 million references to scientific and technical energy related literature. In addition to nuclear science and technology and basic scientific studies in biology, chemistry, engineering, geology, physics, and environment and pollution, the database contains more than one million entries in areas not considered strictly energy fields, including nuclear medicine, computers and mathematical models. The database is updated twice a month and available online from DIALOG (Knight-Ridder) and STN; it is also available on CD-ROM from DIALOG and SilverPlatter.

Energy research abstracts is the successor to *Nuclear science abstracts*, which began publication in 1946 as *Abstracts of classified documents*. The publication changed its name in 1948 to *NSA*, when its scope was widened to include material other than reports from the

USAEC. Five indexes are provided for approaching the contents of each issue of *Energy research abstracts*, namely:

(1) corporate author index;
(2) personal author index;
(3) subject index;
(4) contract number index;
(5) report number index.

The corporate author index lists many well-known establishments around the world, including the Argonne National Laboratory, Illinois; the International Centre for Theoretical Physics, Trieste; the Japan Atomic Energy Research Institute, Tokyo; and the United Kingdom Atomic Energy Authority, Harwell. The subject contents of *ERA* are arranged according to a scheme which provides for 39 first level and 308 second level subject categories; full details of the scheme's scope, definitions and limitations are contained in the report DoE/TIC-4584 *Energy database: subject categories and scope*. The report number index lists items issued from all the major reports series, and one of the largest is the *CONF- series*, which provides information on conference papers and conference proceedings. In addition to *Energy research abstracts*, DoE publishes a number of titles available in the *NTIS energy collection*, and recent titles include *Transportation energy data* (DE94–015723CDG), *Fuel cells* (DE94–004072CDG) and *Heat pumps* (PB95–866281-CDG).

Other energy announcement services

Whilst *ERA* is the principal United States announcement service on energy matters as reflected in reports, many other publications also deal with the subject. For example, *Government reports announcements and index (GRA&I)* includes categories for energy and for nuclear science and technology, and *Scientific and technical aerospace reports (STAR)* caters for documents concerned with energy under the main heading 44: Energy production and conversion, which includes specific energy conversion systems, e.g. fuel cells; global sources of energy; geophysical conversion; and wind power. In addition, the reader is referred to a number of other headings, including 07: Aircraft propulsion and power; 20: Spacecraft propulsion and power; and 28: Propellants and fuels.

In Europe, access to energy related literature published mainly in Germany and German-speaking countries can be achieved by consulting *Energie*, a database produced by FIZ, Karlsruhe. The coverage of the database runs from 1976 to date, and citations are also processed for the *Energy* and *INIS databases*, with 70 per cent of the entries containing both English and German descriptors.

International developments

Whereas *Nuclear science abstracts* been broadened in scope into *Energy research abstracts*, the International Nuclear Information Service (INIS), operated by the International Atomic Energy Agency (IAEA) in Vienna has continued to concern itself primarily with the literature of nuclear science and technology. The exception is economic and environmental aspects of non-nuclear energy sources, which have been monitored since 1992. The subject fields which INIS covers are: nuclear power; nuclear safety; materials of nuclear interest; environmental aspects of nuclear and non-nuclear energy sources; economic aspects of nuclear and non-nuclear energy sources; safeguards and non-proliferation; nuclear applications; radiation protection; nuclear aspects of physics; nuclear aspects of chemistry; and legal matters. INIS is a cooperative decentralized information system set up by the IAEA following the approval of the Board of Governors in 1969. The purpose of INIS is to create a database identifying publications relating to nuclear science and its peaceful applications. Member states and cooperating international organizations (of which there are 94 and 17 respectively) scan the scientific and technical literature published within their boundaries or by them, select from it those items which fall within the subject scope noted above, and process the data according to agreed standards and rules. Document descriptions, abstracts and, in certain cases, full texts are then submitted to the IAEA headquarters, where all the incoming information is merged and the INIS output products compiled. The INIS titles include:

(1) *INIS atomindex*, a semi-monthly announcement and abstracting journal with a main entry section (bibliographic description and abstract) and a number of indexes. As from 1997, the first issue of each volume will include information on the key journals of each inputting centre;

(2) *INIS non-conventional (grey) literature on microfiche*, a service which supplies on request microfiche copies of most of the non-conventional literature (NCL) announced in the INIS output products, with non-conventional literature taken to comprise scientific and technical reports, patent documents, pre-conference papers, and non-commercially published theses. In 1995, the INIS Clearinghouse completed a pilot imaging project to demonstrate the feasibility of electronic storage for NCL on CD-ROM. During the project, six CD-ROMs containing over 50 000 pages of non-conventional literature and related bibliographic data, plus retrieval software, were produced, and as a result INIS is planning for a CD-ROM system which will eventually replace microfiche for the dissemination of NCL;

(3) The *INIS database*, the online service of the IAEA which dates from 1976 and contains over 1.6 million records. The database is updated semi-monthly, with an average annual increase of 80 000 records.

(4) *INIS reference series*, a number of basic manuals including *INIS-1 Guide to bibliographic description, INIS-6 Authority list for corporate entries and report number prefixes* (both of which form the basis for the SIGLE descriptive cataloguing rules), and *INIS-13 Thesaurus*. Full details of INIS products and services are provided in the booklet *Presenting INIS* (GEN/PUB/013.Rev 9).

British sources of energy information

Until 1945, responsibility for atomic energy matters in the United Kingdom rested with the Department of Scientific and Industrial Research (DSIR). In that year, the responsibility was transferred to the Ministry of Supply. Shortly afterwards, the Atomic Energy Research Establishment (AERE) was set up at Harwell, and was complemented by the Production Group at Risley and by the Weapons Group formed in 1947.

Initially, the various establishments founded under the Atomic Energy Act of 1946 were fully engaged in defence commitments, but gradually emphasis moved to the peaceful uses of atomic energy. 1954 saw the passing of the Atomic Energy Authority Act, by which piece of legislation the United Kingdom Atomic Energy Authority (UKAEA) was created as a public board. The subsequent developments of UKAEA in its task of carrying out the research and development necessary to ensure that nuclear power is economic, safe and environmentally friendly; the creation of British Nuclear Fuel plc to provide nuclear fuel services covering the whole fuel cycle; and the acts of privatization such as the formation of British Energy, the nuclear plant operator, are of continuing interest but are outside the scope of this work.

However, as far as scientific and technical information is concerned, the UKAEA, in accordance with the provisions of the Atomic Energy Acts, has made publicly available the results of its work, except where such results need to be protected on the grounds of national or commercial security. Access to abstracts of UKAEA documents can be gained through *Energy research abstracts* and *INIS atomindex*. In addition, UKAEA has for many years published its own regular lists of reports and other documents, wherein it identifies items as:

(1) those deposited at the British Library Document Supply Centre;
(2) those on sale through HMSO and TSO;
(3) those which have appeared in the literature, as for example journal articles.

The items sent to BLDSC are noted in *British reports translations and theses (BRTT)* under section 10: Energy and power, where in particular the following subheadings are used – 10E fission fuels; 100 nuclear power plants; and 10P nuclear reactor technology.

The broader aspects of the energy scene within the United Kingdom used to be coordinated by the Department of Energy but, as has been noted above, responsibility for policies for all forms of energy, including its efficient use and the development of new resources, and the government's relations with the energy industry generally, now reside with the Department of Trade and Industry. Reports on non-nuclear energy topics are also regularly announced in *BRTT*, as for example *Carbon products from coal* (92–09–10A-001), *Energy efficiency in schools* (92–09–10S-001) and *Energy efficiency in hotels: a guide to cost-effective lighting* (96–11–10S-006). A further source of information is the British Library's current awareness bulletin *Focus on British environmental sciences research*, which highlights the latest reports added to the National Reports Centre, plus translations and theses. Finally, mention must be made of an important guide (Chester, 1986) to selected literature and sources of information on nuclear energy and the nuclear industry held by the British Library, including that on open access at the Science Reference Information Service (SRIS) in London. The guide covers abstracts, bibliographies, periodicals, patents, business literature and directories.

Public concern

As with other major topics covered in this book (especially medicine and education), the question of energy is one which touches individuals in every country of the world, and as a result a new grey literature class is emerging which originates not with the traditional government departments, agencies and laboratories, but with pressure groups and institutions on the fringes of or quite outside the core activities of the energy industry. Thus, the daily press regularly carries details of expressions of concern and even outrage over energy issues of the day. Political decisions such as those affecting the future of coal mines, and occurrences of major environmental disasters like the foundering of an oil tanker off a sensitive coastline each generate a host of reports, both official and unofficial, which all go to swell the volume of grey literature. Plainly, they cannot be ignored in the short term and may also have a long term reference value.

References

Anthony, L.J. (ed.) (1988) *Information sources in energy technology*. London: Butterworths

Berkovitch, I. (1996) *Energy sources and policy: an overview and guide to the literature*. London: British Library

Chester, K. (1986) *Nuclear energy and the nuclear industry*, 2nd edn. Boston Spa: British Library

Crewe, A. and Katz, J.J. (1969) *Nuclear research*. New York: Dover

Eagle, S. and Deschamps, J. (eds.) (1997) *Information sources in environmental protection*. London: Bowker-Saur

Fermi, L. (1954) *Atoms in the family – my life with Enrico Fermi*. Chicago: University of Chicago Press

Groves, L.R. (1962) *Now it can be told – the story of the Manhattan Project*. London: André Deutsch

Harris, J. (1996) Nuclear engineering. In Mildren, K.W. and Hicks, P.J. *Information sources in engineering*, 3rd edn. London: Bowker-Saur

Hewlett, R.G. and Anderson, O.E. (1962) *The new world 1939/1946* (Volume 1 of the *History of the United States Atomic Energy Commission)*. University Park, PA: Pennsylvania State University Press

Hewlett, R.G. and Duncan, F. (1969) *Atomic shield 1947–1952* (Volume 2 of the *History of the United States Atomic Energy Commission)*. University Park, PA: Pennsylvania State University Press

McMullen, J.T. (1996) *Energy technology*. In Mildren, K.W. and Hicks, P.J. *op.cit.*

Murray, R.L. (1988) *Nuclear energy*, 3rd edn. Oxford: Pergamon Press (see especially Chapter 23: Laws regulations and organizations)

Physical review, (1939), **55**, 416–418, 511–512, 797–800; **56**, 284–286, 426–450

Smyth, H.D. (1945) *Atomic energy for military purposes*. New York: Princeton University Press

CHAPTER TWELVE

Science and Technology

Introduction

The literature of science and technology is large by any standard, and the proportion of it represented by grey literature, in particular by reports, is quite significant. Indeed, impressive though the coverage of the conventional literature is through long-established and all-embracing services such as *Chemical abstracts*, the *INSPEC* abstracts journals, and *Engineering index*, no survey of the literature of a particular branch of science or technology can be considered complete or comprehensive unless due account is taken of the sources of information contained in the grey literature.

The reasons why abstracting and indexing services covering science and technology have flourished and expanded over the years is because they are vital for the success of research and development programmes which in turn are the driving force of science based technologies and ultimately affect and influence the progress of whole sectors of industry. Keys to the literature of science and technology have long been recognized as an important component in both the academic and industrial fields of endeavour, and consequently sufficient funding has been forthcoming to enable information services to be organized and sold on a self-sustaining basis.

At first, when reports and other types of grey literature began appearing on the scene, their treatment was somewhat haphazard, but as the realization of their value as sources of information complementary to the conventional literature files grew, efforts were made to coordinate the accessioning and announcement of such documents. The result has been a measure of control equal to or, in some cases, even better than that of the published literature, and an attempt will be made here to indicate some of the more important guides currently available. The field of science and technology is a widely defined one embracing a great number of separate disciplines, and noticeable, at least in terms of reports literature, for the significant presence of aerospace and energy, topics which are dealt with separately in Chapters 6 and 11 respectively. The life sciences, too, account for a great deal of the output of the conventional publishing services, but less so in the case of grey literature; they are considered in Chapter 7.

The actual effectiveness of the use of grey literature in the industrial sector is not easy to gauge. It is one thing for an agency or a government department to specify and promote research and development in a particular project or addressed to a particular problem; terms of reference are agreed, a sum of expenditure is named and a budget drawn up, and the work executed and reported upon. It is quite another matter to judge how well companies or indeed sectors of industry adapt the results of the same piece of research and development work for their own purposes. The outcome may indeed be an addition to the general sum of knowledge, but whether it is exploitable further is a completely different question. This problem, as has already been observed, greatly exercises the National Aeronautics and Space Administration, which continues to make great attempts to stimulate technology transfer over a wide range of industries.

In surveying science and technology to consider its grey literature aspects, it is important to acknowledge the broader picture and draw attention to the growing list of relevant titles in the series of which this work forms part. The key volumes are *Information sources in physics* (Shaw, 1994), *Information sources in engineering* (Mildren and Hicks, 1996), *Information sources in chemistry* (Bottle and Rowland, 1993), *Information sources in polymers and plastics* (Adkins, 1989), *Information sources in patents* (Auger, 1992), and *Information sources in metallic materials* (Patten, 1990). It is also useful to mention a work of a comprehensive but somewhat different nature, *Information sources in science and technology* (Parker and Turley, 1986), which is a practical guide to traditional searching techniques and online usage, and which constitutes a reference work aimed at both the organizers and the users of technical information. Finally, since much of the grey literature in this area is of United States origin, readers will find invaluable the work by Aluri and Robinson (1983), *A guide to US government scientific and technical resources*, an encyclopaedic compilation covering among other topics technical reports, scientific translations, indexes and abstracts, databases and information analysis centres.

National Technical Information Service

In the United States, the organization which dominates the grey literature scene, especially insofar as it concerns reports, is the National Technical Information Service (NTIS), an agency of the US Department of Commerce. NTIS is the central source for the public sale of US government sponsored research, development and engineering reports, and for the sale of foreign technical reports and other analyses prepared by national and local government agencies and their contractors and grantees. In order to appreciate the present day prominence and influence of NTIS, it is worth recounting how it all began. The body of public law that serves as the 'charter' for NTIS essentially provides the agency with a mandate to disseminate scientific, technical and engineering information to the American public. Details of the relevant Executive Orders and Public Laws which apply are conveniently summarized in Appendix A of the work by McClure and co-workers (1986), an important study already mentioned in Chapter 2.

The key dates in the history of NTIS, together with the names of the various agencies involved, are:

(1) 1945–46 Office of the Publication Board (PB – established by Executive Order 9568:1945);
(2) 1946–65 Office of Technical Services (OTS – established by Executive Order 9809:9146);
(3) 1965–69 Clearinghouse for Federal Scientific and Technical Information (CFSTI, an agency placed under the National Bureau of Standards);
(4) 1970 to date National Technical Information Service (NTIS – CFSTI was abolished and its functions transferred to the new body).

Today, in terms of size, the NTIS activities are enormous; the collection comprises around two and half million titles, several hundred thousand of which contain foreign technology or foreign marketing information. All the titles are for sale, mostly as copies from microform masters, but many titles in frequent demand are held as shelf stock. NTIS adds approximately 75 000 new works to its collection annually, with nearly one third of the new additions coming from foreign sources, including Japan, eastern Europe and Russia. An example of overseas participation in NTIS has been described by Fazio (1995), who reviews the role of the library of the National Research Council (CNR) Rome, in the promotion of grey literature.

The NTIS has for some years made a practice of naming its best sellers, and currently the top three titles are: the *Statistical abstract of the United States* (PB95–965801-CDG), compiled by the Bureau of census; the *World fact book* (PB95–928009-CDG), compiled by the

Central Intelligence Agency; and the *United States government manual* (PB96–114012-CDG), compiled by the National Archives and Records Administration. Some of the major services based on the NTIS collections are as follows:

(1) *NTIS alerts* (formerly *NTIS abstract newsletters)*, a service available in two forms, namely the *Custom NTIS alert*, in which a set of topical search criteria is run twice monthly against the newest NTIS studies and research results, and the *Prepackaged NTIS alert*, in which R&D reports in 26 subject areas are summarized on a twice monthly basis and are available on subscription;

(2) Online access to the records in the *NTIS bibliographic database*, which dates from 1964 and is now available through a number of hosts, including DataStar, Dialog, ESA, Ovid Technologies, Questel-Orbit, SilverPlatter and STN;

(3) *NTIS published searches*, produced jointly by NTIS itself and NERAC Inc. (see below), which are bibliographies containing 50 to 250 latest abstracts from a range of databases which extend from the *Aerospace database* to *World surface coatings abstracts (WSCA)*; NTIS also issues a *Master catalog* to the *Published search* files (PR-186). NERAC Inc. is a problem solving and transfer centre in Tolland, Connecticut, which prepares and produces the *Published search* bibliographies and mails them to NTIS customers in response to orders;

(4) *Selected research in microfiche (SRIM)*, a facility whereby clients can receive full texts of reports in microfiche distributed according to a previously agreed user profile as selected from 350 subject areas;

(5) *Federal research in progress (FEDRIP)*, a database providing a means of access to summaries of over 150 000 US government funded research projects currently in progress, and compiled from a number of government sources including the Department of Energy, the National Aeronautics and Space Administration, and the Small Business Innovation Research Program.

Until quite recently, NTIS relied upon voluntary transfers by federal agencies to ensure the comprehensiveness of its collection. However, since the passing of the American Technology Pre-eminence Act (ATPA – Public Law 102–245) in 1991, the wealth of information available to NTIS has grown considerably. The ATPA requires all federal agencies to submit their federally-funded scientific technical and engineering information to NTIS within 15 days of the date of the product being made publicly available. Items of information so received are entered into the *NTIS bibliographic database* and become a permanent part of the NTIS collection. Over 200 US government agencies now contribute to NTIS, which also has access to certain White House documents as well.

Other services offered by NTIS, and more information on those noted above, can be found in the NTIS *Catalog of products and services*, issued annually and obtainable on request as PR-827. General descriptions and evaluations of the work of NTIS can be found in the comprehensive survey by Caponio and MacEoin (1991), which places special stress on NTIS liaison with librarians and information specialists; and in the study by Khan and Carrol (1992), which emphasizes the role of the report in the transfer of scientific and technological information.

Government reports announcements and index (GRA&I)

The principal vehicle for the notification by NTIS of the reports and other grey literature items as they are added to the main collection is *Government reports announcements and index (GRA&I)*, which has for many years been published twice a month with details of around 3000 bibliographic citations in each issue. It is completed by a comprehensive annual index, and like NTIS itself, has undergone several identity changes, which are summarized in Table 12.1.

GRA&I is a comprehensive publication covering an extremely wide range of topics, and is particularly aimed at libraries and information centres. Until December 1986, *GRA&I* used the NTIS subject category and subcategory structure endorsed in 1964 by the Committee on Scientific and Technical Information (COSATI) of the Federal Council for Science and Technology, described in report AD-612 200. The use of COSATI corporate author headings is described report in PB-198

Table 12.1 GR&I and antecedents

Dates	Agency	Title	Frequency
1946-49	OTS[1]	*Bibliography of scientific and industrial reports*	Weekly
1949-54	OTS	*Bibliography of technical reports*	Monthly
1955-61	OTS	*US government research reports*	Monthly
1961-64	OTS	*US government research reports*	2 per month
1965-69	CFSTI[2]	*US government research and development reports (with separate index)*	2 per month
1970-71	NTIS[3]	*US government research and development reports (with separate index)*	2 per month
1971-75	NTIS	*Government reports announcements (GRA) indexed by Government reports index (GRI)*	2 per month
1975 to date	NTIS	*Government reports announcements and index (GRA&I)*	2 per month

Notes [1] OTS = Office of Technical Services;
[2] CFSTI = Clearinghouse for Federal Scientific and Technical Information;
[3] NTIS = National Technical Information Service

275. From January 1987 onwards, *GRA&I* began to use the NTIS subject category and subcategory structure version which can also be used for online searching, and which is described in PR-832 *NTIS subject category descriptions*. The original COSATI list had 22 numbered categories; the current list has 38, which are no longer numbered. Details of the NTIS subject classification system (past and present) are given in report PB-270 575.

Each issue of *GRA&I* is divided into:

1   Reports announcements;
2   Keyword index KW;
3   Personal author index PA;
4   Corporate author index CA;
5   Contract/grant number index CG;
6   NTIS order/report number index OR;
7   Price codes.

Each bibliographical entry in each issue of *GRA&I* is allocated an abstract number, and the documents themselves are usually identified by one of the following major report series codes:

(1) AD-A, representing unlimited, unclassified documents originating with the Department of Defense (AD once stood for ASTIA document, and ASTIA in turn was the Armed Services Technical Information Agency);
(2) DE, representing documents cited in *Energy research abstracts*, compiled by the Department of Energy;
(3) N97, representing NASA documents;
(4) PB97, representing documents processed by NTIS itself (PB originally meant Publication Board, the ultimate forerunner of NTIS).

Many other identifiers are included in the indexes, especially CONF- (conference papers); EPA (Environmental Protection Agency); ISBN (International Standard Book Number); and TIB (Technische Informationsbibliothek Hannover).

The richness and diversity of the grey literature covered so thoroughly in *GRA&I* can be assessed in two ways. Firstly, there is the vast range of document formats which are regularly cited, of which the following list is just a sample:

Annual reports	Journal papers
Bibliographies	Leaflets
Briefing reports	Masters' theses
Bulletins	Memorandum reports
Conference papers	Patent applications
Contract reports	Progress reports
Data files	Quarterly reports
Doctoral theses	Research reports

Draft reports	Summary reports
Final reports	Translations
Granted patents	USGPO publications
Internal reports	Working papers

Secondly, there is the sheer spread of the NTIS subject category and subcategory structure, such that an A to Z sample of indexing terms from a typical issue can read:

A	Absenteeism	N	Nickel
B	Benzene	O	Optical images
C	Circuit breakers	P	Potable water
D	Dosimeters	Q	Quartz
E	Energy policy	R	Rockets
F	Families	S	Sapphire
G	Gradients	T	Taxes
H	Hazardous materials	U	Unsteady state
I	Incinerators	V	Vision
J	Jamaica	W	War games
K	Kidney calculi	X	X-rays
L	Laser targets	Y	Yukon Territory
M	Meetings	Z	Zinc

There is something of a penalty to be paid for such a wide coverage, and that is the significant amount of overlap among the major bibliographic sources for grey literature. For example, a study by Copeland (1981) found that *Government reports announcements and index* covered 94.3 per cent of the technical reports that were indexed in *STAR* and 78.8 of those that were indexed in *ERA*. Conversely, it was concluded that about 41 per cent and 21 per cent of report entries in *GRA&I* turned up in *STAR* and *ERA*, respectively. Whilst at first sight such duplication might appear undesirable as a misuse of effort, some authorities hold the view that the repetition of entries provides an improved access to the technical reports in question – especially if the user is aware of the extent of the duplication in the announcement services.

NTIS has come to the conclusion that the market for its traditional paper and fiche products, whilst admittedly shrinking, will remain the organization's baseline. At the same time, NTIS is taking the lead in making a major shift from paper to electronic publishing when it comes to government information dissemination, both domestically and internationally. Indeed, the newsletter *NTIS newsline*, published in conventional format for nearly 50 issues, is now to be found only on the NTIS World Wide Web Home Page. Significantly, too, in 1996 plans were announced for the publication of *GRA&I* in printed paper format to be discontinued; henceforth, the venerable announcement service will be available in electronic format only, a step also taken by NASA in respect of *Scientific and technical aerospace reports* (see Chapter 6).

Sales of NTIS documents and other products, as distinct from the lending and reference facilities offered by major national and public libraries, are handled by a worldwide network of official distributors. In the case of the United Kingdom, the official distributors, known as the Official UK Managing Dealer, are Microinfo Limited, a firm of publishers and distributors of a range of specialist information. Microinfo was founded in 1970 and in 1976 was awarded a full official agency for all the products and services of NTIS, details of which are regularly summarized in Microinfo catalogues and sales leaflets

RAND

Among the many United States organizations which issue reports and other items of unpublished literature on any scale, the Rand Corporation is noteworthy in that it has produced its own announcement journal, *Selected RAND abstracts*. The Rand Corporation (the name derives from a contraction of *Research and Development*) is an independent non-profit organization engaged in scientific research and analysis. The Corporation conducts studies in the public interest supported by the US government, by local and state governments, its own funds derived from earned fees, and by private foundations. The work involves most of the major disciplines in the physical, social and biological sciences, with the emphasis on their application to problems of policy and planning in domestic and foreign affairs. The output of generally available RAND titles is not large – typically 250 to 300 items each year, which in addition to being announced by RAND itself are regularly noted by a number of major agencies, including NTIS, NASA and ERIC. The subject coverage is extremely wide, and a typical report title is *The political evolution of anti-smoking legislation* (R-4152-UCOP). Further information may be found in the regular issues of the *RAND research review*, which gives details of programmes on matters affecting the security and domestic welfare of the United States.

Defence

Documents originating from defence sources, especially the aerospace and energy fields, as has already been noted, have always been a major feature of grey literature, and in science and technology in general one of the major sources of such material in the United States is the Defense Technical Information Center (DTIC), described by Lahr (1981) as 'but one part of a complex bureaucracy which comprises the DoD (Department of Defense) technical information program'. The DTIC, which serves the DoD and its contractors as well as other US government agencies and their contractors, can trace its origins back to an effort originally established in Great Britain, when a system was set up in London in 1945 which was designed to process confiscated German technical documents. By 1951, the United States documentation centres involved – namely the Navy Research Section of the Library of Congress and the Central Air Documents Office at Dayton, Ohio, operated by the US Air Force – had been consolidated into an organization already alluded to, the Armed Services Technical Information Agency (ASTIA). In 1963, the name of the Agency was changed to the Defense Documentation Center (DDC) which eventually became the present-day Defense Technical Information Center (DTIC).

Whereas NTIS deals with unclassified and unlimited distribution reports and other documents freely accessible to the general public, DTIC handles classified and limited distribution reports and other documents available only to a limited clientele. As a result, access to DTIC is strictly restricted, and requires prior registration and the certification of a 'need to know'.

Details of the products and services offered by the DTIC to qualified users are quite frequently mentioned in documents abstracted *GRA&I*, as for example *How to get it: a guide to defense-related information sources (revision) 1995*. The *Guide* is available from NTIS as report AD-A256 150 CDG and covers the identification and acquisition of US government sponsored or published documents, maps, patents, standards and other defence related sources. Another useful reference tool is the *Directory of organizational technical report acronym codes (DOTRAC)*, a guide to acronyms as assigned by the DTIC, which is in three parts, namely alphabetically by acronym, by the full name of each organization, and numerically by corporate author (source header) code (DOTRAC 1991).

Although DTIC does not serve the general public directly, the scientific and technical community does have access to reports and other documents having no security or distribution restrictions. Such items are sent to NTIS where they are announced through *GRA&I*. This policy has some interesting consequences, and indeed is not without its critics. Thus, the Director of the US Defense Technology Security Administration is quoted as saying 'During the Carter administration a system was set up within the US government to get rid of excess classified documents. The US Defense department was sending everything but classified information to NTIS. But there was an automatic seven year 'clock' on declassification. When a document was seven years old it was automatically declassified. The documents were about everything, tank warfare, missile fuels, electro-optics, advanced computer databases, radar absorption. They were all dumped at NTIS when they were automatically declassified. So the Soviets became the best customers of NTIS', (Fisk, 1988).

Presentations on the work of DTIC are made at the *Annual users' conferences*, the proceedings of which are regularly notified in *GRA&I* (see, for instance report AD-A-166–250/1/GAR).A recent development has been the preparation of a CD-ROM entitled *Defense library on disc*, which combines details of the holdings of the Pentagon Library and the National Defense University Library, and gives access to 210 000 records covering international affairs, political science, military affairs, management logistics, computer science and US law. The CD-ROM is available by subscription through NTIS.

The United Kingdom and European scene

In the United Kingdom and in Europe, provision for the handling of science and technology reports and other grey literature has been established on a much more modest scale than in the United States. At one time, the British Ministry of Technology (MinTech) operated a Technology Reports Centre (TRC) which processed and made available exploitable and unpublished research and development reports arising from United Kingdom government programmes and those of overseas governments. The TRC held the majority of non-classified technical reports produced in government research establishments, and in some cases the reports series extended as far back as 1940. The Centre's holdings were particularly strong in electronics, aeronautics, materials technology and mechanical, electrical and industrial engineering. The Centre also received the openly available publications provided by NTIS and NASA, plus many other documents from US and overseas organizations supplied under exchange agreements.

TRC had a number of methods for publicizing the contents of its collection, the most important of which was the twice-monthly announcement service *R&D abstracts*, a publication arranged according to the COSATI scheme. Documents cited therein could be obtained as paper copies or as microfiche, and a set of comprehensive indexes was compiled on a half-yearly basis. However, as the collection at TRC developed and grew in size it became apparent that there was an increasing overlap between the reports processing activities of the Centre and those of the British Library. A decision was therefore taken that TRC should withdraw from reports handling and instead place reliance on the British Library, with its far greater resources and economies of scale. As a result, TRC's stock of T-numbered reports for the years 1972–81 was transferred from TRC's premises at St Mary Cray, Kent, to the British Library at Boston Spa, where the documents were made available to registered users. An exception was made for technical reports and other documents with a defence classification, which continued to receive special treatment, and which today are handled by the Glasgow-based

Defence Research Information Centre (DRIC), described below. The decision to end TRC's report handling activities also meant the demise of *R&D abstracts*, the final issue of which appeared on 15 December 1981. The file still retains some reference value, not least because the abstracts were very full and informative.

Nowadays, the main announcement journal for British reports and other grey literature material with an unrestricted availability is *British reports translations and theses (BRTT)*, a bibliography published monthly (with indexes which cumulate annually) by the British Library Document Supply Centre (BLDSC). *BRTT* is provided as an aid to practitioners and researchers who need to trace information contained in reports – including market research material – produced by British government bodies, industry, universities and other learned bodies and held in the National reports Collection. In addition, as the title indicates, *BRTT* covers British translations and theses. The material listed in *BRTT* also appears in the System for Information on Grey Literature in Europe (SIGLE) database, discussed in Chapter 2.

At one time, entries in *BRTT* were arranged by a system based on the COSATI classification scheme, available as report AD-612 200. In 1993, mention of COSATI was discontinued and replaced by a reference to the SIGLE classification scheme, a modified version of the original COSATI headings. In considering the use of the new headings, it is noticeable how the material listed under section 05: Humanities, psychology and social sciences has increased in recent years, as the concept of grey literature has been widened outside its original base of science and technology. Thus it is now possible to find in *BRTT* citations relating to a colloquium on Homer (93–01–05L-044), a lecture on Old Testament ethics (93–01–05H-018), the sociology of time-space compression (96–11–05R-002) and the Roman art of war as revealed by a study of Roman military writers (96–11–05E-010). Other categories too have seen considerable enlargement of coverage, especially since 1990, with an increase in the amount of attention devoted to the environment, economics computer science and food technology.

Unlike *R&D abstracts*, the entries in *BRTT* carry no summaries but simply a basic amount of bibliographical detail and, where considered necessary, an augmented title. For example, the report called *The making of Britain's best factories* (96–11–05X-085) is augmented by the words [UK manufacturing strategy]. *BRTT* is supplemented by four current awareness bulletins in the *Focus* series which cover British biological and medical sciences, business and management sciences, environmental sciences, and engineering and computer sciences.

Defence Research Information Centre

Whereas with the closure of the Technical Reports Centre and the ending of the publication of *R&D abstracts*, much of British industry was left to its own devices in the identification and monitoring of grey literature in the practical application of *R&D*-based science and technology, in the case of companies engaged in the defence industries, and so likely to be working on government contracts, practical help continued in the shape of the services provided by the Defence Research Information Centre (DRIC), a part of the Defence Evaluation and Research Agency (DERA). The basic role of DRIC is to disseminate scientific and technical information, especially that contained in unpublished reports, to the UK defence community, which is defined as Ministry of Defence branches and establishments, armed services units, military colleges, and UK firms and organizations working on government defence contracts. DRIC adds to its collection in three main ways:

(1) direct acquisition of documents from the originators (MoD branches, establishments and contractors);
(2) acquisition of documents from overseas defence information centres, in particular from the United States, Canada and Australia;
(3) acquisition of documents as a result of requests for specific titles.

DRIC has more than 600 000 reports dating back to World war II and acts as the focal point for the exchange of scientific and technical (S & T) defence reports with other countries. The centre is able to draw on resources and skills built up during the 75 years or so since its inception as Admiralty Technical Records. DRIC has its own announcement service *DRIC abstracts*, published monthly in two editions:

(1) MoD edition – with the protective marking *Confidential* and a quarterly *Secret* supplement;
(2) Contractors' edition – with the protective marking *Restricted*.

The DRIC also produces two bulletins concerned with overseas reports, namely:

FDRB – Foreign defence reports bulletin, which contains citations listing recent reports acquired by the Canadian, German, Swedish and Swiss departments of defence, and by the SHAPE Technical Centre in the Netherlands; report listings from the French and Australian departments of defence are included when available;
USDRB – US defence reports bulletin, which gives details of *Unlimited* and *Limited* distribution reports supplied to DRIC by its equivalent organization in the United States.

DRIC issues a regularly updated set of comprehensive information leaflets which describe its services; the set is available on request from the Glasgow headquarters.

Euroabstracts

A bi-monthly announcement service published by the European Commission is *Euroabstracts*, an abstracting journal containing details of documents resulting from Research and Technological Development (RTD) programmes financed wholly or partly from the budget of the European Communities. Cited publications include:

(1) *EUR reports* – scientific and technical studies, monographs, proceedings of conferences, workshops, and contractors' meetings;
(2) Other reports and documents produced by the Commission relating to Community RTD activities;
(3) Articles and papers relevant to the Commission's research activities.

Euroabstracts, which contains about 71 000 references from 1963 onwards, is available online as the database *CORDIS RTD publications*, one of nine databases produced by the Community Research and Development Information Service (CORDIS). Elsewhere in Europe, the Scientific and Technical Information Centre of Russia (VNTIC) publishes *R&D abstracts journal* in ten subject areas, and features summaries of reports and theses from the CIS countries.

Summary

As noted at the start of this book, reports literature grew out of the need of scientists and engineers to communicate the results of research and development quickly, cheaply and efficiently, with security controls applied if necessary. Technical reports now constitute a considerable part of the grey literature but paradoxically their role, impact and importance are still the subject of conflicting views. An attempt has been made by McClure (1998) to summarize current options on three fundamental issues:

(1) the use of the technical reports literature by the *R&D* community;
(2) the manner in which the technical report assists the *R&D* community;
(3) an assessment of the value of technical reports.

The paper indicates that further lines of investigation are called for if we are to get an increased knowledge of the subject, and indeed the ongoing NASA/DoD Aerospace Knowledge Diffusion Research Project, noted in Chapter 6 is an example of such activity. In the meantime, the fact remains that in the realm of science and technology, reports form an important medium for furthering the processes of research and innovation, the transfer of scientific and technical information, and the achievement of technology transfer. These objectives have been emphasized as far as Great Britain is concerned by a report commissioned by the Department of Trade and Industry (1993) which notes that companies need to promote innovation to survive, and that it is vital they tap into the *R&D* of customers and suppliers in order to upgrade their own technical competence. One method available, of course, is to make greater use of the grey literature.

Finally, it has to be stressed that the grey literature of science and technology depends heavily on United States agencies and announcement services for its effective dissemination and bibliographical control. European efforts, notably through *British reports translations and theses* and Germany's *Forschungsberichte* are making increasingly important contributions, but users still rely very heavily on *Government reports announcements and index, Scientific and technical aerospace reports*, and *Energy research abstracts*.

References

Adkins, R.T. (ed.) (1989) *Information sources in polymers and plastics*. London: Bowker-Saur

Aluri, R. and Robinson, J.S. (1983) *A guide to US government scientific and technical resources*. Littleton, Colorado: Libraries Unlimited.

Auger, C.P. (ed.)(1992) *Information sources in patents*. London: Bowker-Saur

Bottle, R.T. and Rowland, J.F.B. (eds.) (1993) *Information sources in chemistry*, 4th edn. London: Bowker-Saur

Caponio, J.F. and MacEoin, D.A. (1991) The National Technical Information Service: working to strengthen US information sources. *Reference librarian*, **32**, 217–227

Copeland, S. (1981) Three technical report printed indexes: a comparative study. *Science and technology libraries*, **1**, Summer, 41–53

Directory of organizational technical report acronym codes (DOTRAC) (1991) Report AD-A-237 000.

Department of Trade and Industry (1993) Innovate or bust. *Guardian*, 28 January 1993

Fazio, A. (1995) Organization and development of a national centre for coordinating and disseminating scientific, technological and industrial information in Italy. In Farace, D.J. (ed.) (1996) *Proceedings of the second international conference on grey literature, Washington, 1995*. Amsterdam: TransAtlantic Publishing

Fisk, R. (1988) US defence information. *The Times*, 8 October 1988

Khan, A.K. and Carrol, B.C. (1992) Industrial and technological information – who has it and who needs it? The role of the NTIS in acquisition and dissemination. *Journal of library and information science*, **18**, (2) 29–45

Lahr, T.F. (1981) *A study on decreased reporting in the Department of Defense*. Report AD-180 238

McClure, C.R. (1988) The Federal technical report literature: research needs and issues. *Government information quarterly*, **5**, (1) 27–44

McClure, C.R., Hernon, P. and Purcell, G.R. (1986) *Linking the US National technical Information Service with academic and public libraries*. Norwood, N.J.: Ablex Publishing Company

Mildren, K. and Hicks, P. (eds.) (1996) *Information sources in engineering*, 3rd edn. London: Bowker-Saur

Parker, C.C. and Turley, R.V. (1986) *Information sources in science and technology: a practical guide to traditional and online use*, 2nd edn. London: Butterworths

Patten, M.N. (ed.) (1990) *Information sources in metallic materials*. London: Bowker-Saur

Shaw, D. (ed.) (1994) *Information sources in physics*, 3rd edn. London: Bowker-Saur

Keys to Report Series Codes

A constant problem in handling inquiries relating to grey literature is the one presented by the reader who has incomplete details of the documents he or she wishes to see. Quite often, the only real clue is the sequence of numbers or of numbers and letters which make up a report series code. Fortunately, several publications are available which give help in identifying and verifying such codes. Some works attempt to cover as many codes as possible, whilst others concentrate on specific subject areas. All are invaluable in helping to recognize and deal with a range of report codes which are increasingly quoted in the open literature without any form of explanation. The characteristics of the major works of this nature are noted below.

Probably the best known compilation is the *Report series codes dictionary* (Aronson, 1986), the purpose of which is to identify, and provide an association for, most of the codes which have been applied to unpublished reports. The *Dictionary* was originally the result of a spare-time effort on the part of members of the Special Libraries Association (SLA) in the United States, and the first edition appeared in 1962. The second edition, edited by L.E. Godfrey and H.F. Redman, who were also responsible for the first edition, appeared in 1973 and is still available through University Microfilms International. The third edition is the outcome of work by the Commerce, Energy, NASA, Defense Information cataloging Committee, Washington, DC, and now runs to 647 pages. It constitutes a guide to over 20 000 report series codes used in the dissemination of scientific, technical and industrial information by nearly 10 000 corporate authors. The *Dictionary* is arranged in two parts, the first alphabetically by report acronym or other identifier, and the second by organization name, with the corresponding report series codes.

A work which complements the *Report series codes dictionary* is the *Directory of engineering document sources*, first published in 1973 under the editorship of Simonton; an update appeared in 1985, with a slight change in the title from *Directory* to *Dictionary*. Taken together, the two dictionaries represent an excellent general approach to resolving the meanings of reports codes and other problems of origin and identification, and as such form essential reference works in any grey literature collection.

Another important work of reference is the *Alphanumeric reports publications index (ARPI)*, originally devised as an in-house listing to assist British Library staff trace individual series through the maze of primary and secondary report numbers at different locations within the Boston Spa site. The entries include the country of origin, either an expansion of the alphanumeric code or a subject heading and the BLDSC stock location. *ARPI* was first announced in 1992, and the fourth edition (1997) allows access by the alphanumeric designator to reports held in some 12 000 series. BLDSC also compiles a *Dictionary of acronyms* (1995) which lists thousands of acronyms culled from the Centre's conference database, and is updated every two years.

In specific subject areas too, it is often possible to refer to specialized guides. For example, the Office of Scientific and Technical Information of the United States Department of Energy has issued *Report number codes*, a compilation in which Part 1 is alphabetical by report codes followed by the names of the issuing agencies, and Part 2 lists the issuing agencies followed by the assigned report code or codes. Other useful sources are the annual indexes to the various announcement journals such as *Scientific and technical aerospace reports (STAR)* and *Energy research abstracts (ERA)*. Finally, as noted in Chapter 12, the Defense Technical Information Center has published its own *Directory of organizational technical report acronyms – DOTRAC* (1991), a guide to acronyms assigned by the DTIC.

References

Aronson, E.J. (1986) *Report series codes dictionary*, 3rd edn. Detroit: Gale Research

British Library (1997) *Alphanumeric reports publications index*, 4th edn. Boston Spa: BLDSC

British Library (1997) *Dictionary of acronyms*, 3rd edn. Boston Spa: BLDSC

Dictionary of engineering document sources (1985) Santa Ana, California: Global Engineering Documents

Directory of organizational technical report acronym codes (DOTRAC) (1991) Report AD-A-237 000

APPENDIX B

Trade Literature

Trade literature possesses many of the attributes of grey literature, notably largely pamphlet format, controlled distribution and varying standards of bibliographical control. Whether it should be treated as part of the grey literature, in that it should be systematically collected and carefully recorded in databases, is not an easy question to answer. Certainly, trade literature cannot be ignored, for it is an important source of certain types of information, but to admit it as a category alongside other categories such as reports, translations, theses and similar documents presents practical problems which are not readily resolved. On the one hand, there are simply enormous quantities of trade literature issued each year, making any attempt to select and organize it time-consuming and expensive; on the other, hand some trade literature, as will be noted below, eventually becomes of great interest to historians, especially those working in the fields of technology and social studies. Thus the consignment of trade literature to an appendix in this work reflects an uncertainty surrounding the status of the material.

The stated purpose of trade literature is to promote sales and provide information about products and services, functions, it is true, some authorities feel are better carried out in separate documents, but which in reality are often combined. Thus, a typical piece of trade literature will consist of a brochure explaining the advantages of a product or service in terms of its benefits and economics, and secondly stating in the form of a specification a summary of key facts and figures. Usually, trade literature in library collections is filed for its information content and the sales aspect is largely irrelevant. The amount of information provided in trade literature will depend firstly on the competence profile of the likely customer, and secondly on the complexity and nature of the product or service. Typically, the information supplied will include performance data, operating characteristics, and materials properties. In certain areas, notably in electronics, trade literature is highly structured and extends to technical reports on the application of discrete devices and circuits. Usually, each commercial organization produces its own trade literature in accordance with how it sees the need of the moment, and variety rather than uniformity is often a deliberate aim.

Trade literature is often regarded as an extremely valuable, up-to-date and detailed source of technical information, and many methods have been advocated to ensure its full utilization in information centres and libraries. In some cases, trade literature is treated as part of a library's total collection and it is accorded an appropriate (that is to say minimal) amount of cataloguing, indexing and storage facilities. The British Library Science Reference and Information Service (SRIS) has a collection of current trade literature from around 20 000 mainly British companies, covering a broad range of manufacturing industries, comprising:

(1) Product literature (catalogues, data sheets, price lists, etc.);
(2) House journals;
(3) Company annual reports;
(4) Stockbrokers' reports;
(5) Exhibition catalogues.

These types of literature are interfiled and arranged alphabetically by company name. In addition, SRIS houses a collection consisting of trade literature published between 1830 and 1940 from about 7500 firms.

Two characteristics of trade literature which undermine any formal attempt, or indeed any incentive, to organize it into collections similar to other library material, as for example technical reports, are (a) that it is easily and readily available to *bona fide* inquirers direct from companies and organizations only too willing to provide details of their activities in the hope of making a sale, and (b) that it is seen by recipients as ultimately expendable because it has no cover price. Notable exceptions do occur, again, for example, in electronics, where certain categories of trade literature are available on a subscription basis. Even when pieces of trade literature do bear a price on the cover, this fact is often ignored by a manufacturer's representative, who is empowered to supply the documents free of charge at his discretion.

Not surprisingly, therefore, many libraries and information centres prefer to concentrate on collecting and maintaining a good set of buyers' guides and trade directories which point the way to specific products and services, and which leave the actual accumulation and subsequent disposal of the trade literature itself to individuals or to technical departments and commercial offices. Directories and similar publications range from the all-embracing (for example, the *Kompass* series) to the very specific (such as the *Carpet annual* the *Gas industry directory* and *World insurance*), and such publications have been surveyed many times; one of the most recent appraisals is that by Stacey and Shilling (1996).

Attempts to standardize the content and especially the format of trade literature have been made from time to time. The British Standards Institution issued *BS 1311* in 1955 on sizes of manufacturers' trade and technical literature (including recommendations for the contents of catalogues), but its influence appears to have been minimal, and the specification was in fact subsequently withdrawn. At present, there are just two British Standards covering trade literature, namely *BS 4462* on the preparation of technical sales literature for measuring instruments and process control equipment, and *BS 4940* on the presentation of technical information about products and services in the construction industry. Good presentation of trade literature depends on a number of key features:

(1) clear technical descriptions;
(2) high quality illustrations;
(3) integrated layout and design;
(4) adequate document identification.

Whatever methods (if any) for the standardization of trade literature are adopted, and whatever methods are advocated for its handling and exploitation, the prime purpose of trade literature will always be the furtherance of trade rather than the general dissemination of information and the enlargement of knowledge. However, as indicated above, despite this pre-eminent concern with immediacy, the producers of trade literature are in many cases unwittingly contributing to a growing archive of valuable historical material, an aspect which is becoming the focus of increasing attention. Thus Connor (1990) has reviewed changes in trade catalogues in the Los Angeles Public Library over a 75-year period, whilst Ratner (1990) notes that trade catalogues document a number of aspects of American history, and so have a research value material culture, graphic arts, printing history, labour conditions, cultural and social values, the history of technology, and the evolution of industries. Further reference to the importance of trade catalogue collections in California has been made by Kurutz (1996), reporting to the Society for Industrial Archaeology. Hence there is a clear argument for greater consideration to be given to questions concerning the availability, accessibility and preservation of such material, especially in microform or CD-ROM format.

For a fuller insight into the nature, use and value of trade literature, and indeed into where some of the solutions may lie, see especially the work by Wall (1992), and the survey by Ruston (1996), which is particularly strong on progress in non-print sources. The emphasis on non-print sources is significant, since a major development in trade literature in recent years has been the widespread use made by companies large and small of the Internet. Many establishments have set up home pages to display information about the goods they manufacture or supply. The facility is especially useful to potential purchasers investigating broad concepts such as the layout of an office using modular furniture and work stations. The owner of the home page is able to measure the interest generated by the information displayed, since each visit to a site, whether deliberate or accidental, is faithfully logged. In 1997, a survey by the Department of Trade and Industry found that one in six British companies had a web site and that a quarter were on the Internet.

The potential offered by the new medium has inspired a number of works on the construction of a home page, as for instance that by Hobbs (1996). Whether the Internet will oust the traditional trade literature formats remains to be seen – already grumbles are being heard that companies are failing to revise or update their home pages, but then the problems of amendments and additions are the same whatever the chosen distribution system.

References

British Standards Institution (1969) *Guide for the preparation of technical sales literature for measuring instruments and process control equipment.* BS 4462: 1969

British Standards Institution (1994) *Technical information on construction products and services.* BS 4940: 1994, parts 1–3

Connor, B.M. (1990) Trade catalogs in the Los Angeles Public Library. *Science and technology libraries*, **10**, (4) 9–14

Hobbs, L. (1996) *Designing Internet home pages.* London: Made Simple

Kurutz, G.F. (1996) California commercial catalogs from hosiery to hardware on exhibit. *California State Library Foundation bulletin*, **56**, (7) 15–28

Ratner, R.S. (1990) Historical research in trade catalogs. *Science and technology libraries*, **10**, (4) 15–22

Ruston, P.J. (1996) *Product information.* In Mildren, K.W. and Hicks, P.J. (eds.) *Information sources in engineering*, 3rd edn. London: Bowker-Saur

Stacey, M. and Shilling, D. (1996) *Business and company information.* In Lea, P.W. and Day, A. (eds.) *The reference sources handbook*, 4th edn. London: Library Association Publishing

Wall, R. A.(1992) *Engineers' guide to product information: sources and use.* Aldershot: Gower Press

APPENDIX C

Some Points of Contact

Advisory Group for Aerospace Research and Development, AGARD, 7 rue Ancelle, 9220 Neuilly sur Seine, France.

American Institute of Aeronautics and Astronautics, AIAA, 370 L'Enfant Promenade, SW, Washington, DC 20024–2518 USA.

American National Standards Institute, ANSI, 1430 Broadway, New York, NY 10017, USA

American Society of Mechanical Engineers, ASME, 345 E 47 Street, New York, NY 10017, USA.

Association for Information Management, ASLIB, 20–24 Old Street, London EC1V 9AP, UK.

BIOSIS, 2100 Arch Street, Philadelphia, PA 19103–1399, USA.

British Library Automated Information Service, BLAISE, 2 Sheraton Street, London W1V 4BH, UK.

British Library, Document Supply Centre, BLDSC, Boston Spa, Wetherby, West Yorkshire LS23 7BQ, UK.

British Library, Science Reference and Information Service, SRIS, 25 Southampton Buildings, London WC2A 1AW, UK.

British Standards Institution, BSI, Inquiry Section, Linford Wood, Milton Keynes MK 14 6LE, UK.

CAB International, Wallingford, Oxon OX10 8DE, UK.

Cambridge Scientific Abstracts, 7200 Wisconsin Avenue, Bethesda, MD 20814, USA.

Chadwyck-Healey Ltd., The Quorum, Barnwell Road, Cambridge CB5 8SW, UK.

CIMTECH, *see* National Centre for Information Media and Technology.

Defence Evaluation and Research Agency, DERA, Q 101 Building, Farnborough, Hampshire GU14 6TD, UK.

Defence Research Information Centre, DRIC, Kentigern House, 65 Brown Street, Glasgow G2 8EX, UK.

Defense Technical Information Center, DTIC, Cameron Station, Alexandria, VA 22304–6145, USA.

Deutsche Forschungs- und Versuchsanstalt fur Luft- und Raumfahrt, DFVLR, Postfach 90 60 58 D-5000 Koln, Germany.

Deutsches Institut für Medizinische Dokumentation und Information DIMDI, Weisshausstr. 27, D-50939 Koln, Germany.

Educational Resources Information Center, ERIC, US Department of Education, Washington, DC 20208–5720, USA.

Engineering Information Inc. Ei, 1 Castle Point Terrace, Hoboken, NJ 07030, USA.

Euromonitor, 60–61 Britton Street, London EC1M 5NA, UK.

European Association for Grey Literature Exploitation, EAGLE, PO Box 90407, NL-2509 LK, Den Haag, Netherlands.

EUR-OP, Office 172, 2 rue Mercier, L-2985 Luxembourg.

European Documentation and Information System for Education, EUDISED, Council of Europe, 67075 Strasbourg Cedex, France.

European Space Agency, ESA-IRS ESRIN, Via Galileo, I-00044 Frascati, Italy.

Eusidic, the European Association of Information Services, 37 Val St Andre, PO Box 1416, L-1014 Luxembourg.

Fachinformationszentrum, FIZ, Karlsruhe, Postfach 24 65, D-76012 Karlsruhe, Germany.

FOI Services Inc, 11 Firstfield Road, Gaithersburg, MD 20878-1703, USA.

Food and Agriculture Organisation, FAO, Via delle Termi di Caracalla, 00100 Rome, Italy.

Grey Net, *see* TransAtlantic.

Institut de l'Information Scientifique et Technique, INIST, Chateau de Monlet, 54514 Vandoeuvre-les-Nancy, France.

International Nuclear Information System, INIS, International Atomic Energy Agency, PO Box 100, A-1400 Vienna, Austria.

International Translations Centre, ITC, Schuttersveld 2, 2611 WE Delft, Netherlands.

Japan Information Center of Science and Technology, JICST, 5–2, Nagatacho 2 Chome, Chiyoda-Ku, Tokyo 100, Japan.

Knight-Ridder Information Inc., 2440 W. El Camino Real, Mountain View, CA 94040, USA.

Microinfo Ltd, PO Box 3, Omega Park, Alton, Hampshire GU34 2PG, UK.

National Aeronautics and Space Administration, NASA, Center for AeroSpace Information, 800 Elkridge Landing Road, Linthicum Heights, MD 21090–2943, USA.

National Agricultural Library, NAL, US Department of Agriculture, Beltsville, MD 20705, USA.

National Centre for Information Media and Technology, CIMTECH, University of Hertfordshire, Hatfield AL10 9AB, UK.

National Technical Information Service, NTIS, US Department of Commerce, Technology Administration, Springfield, VA 22161, USA.

Questel-Orbit, Le Capitole, 55 Avenue des Champs Pierreux, 92029 Nanterre Cedex, France.

RAND Corporation, 1700 Main Street, Box 2138, Santa Monica, CA 90407–2138, USA.

SilverPlatter Information Inc, 100 River Ridge Drive, Norwood, MA 02062–5043, USA.

Society of Automotive Engineers, SAE, European Office, 27/29 Knowl Place, Wilbury Way, Hitchin, Herts. SG4 OSX, UK.

Special Libraries Association, SLA, 1700 Eighteenth Street NW, Washington, DC 20009–2514, USA.

The Stationery Office, TSO, St Crispins, Norwich NR3 1PD, UK.

STN International, c/o FIZ Karlsruhe, PO Box 2465 D-76012 Karlsruhe, Germany.

TransAtlantic Grey Net, Koninginneweg 201–1075, CR Amsterdam, Netherlands.

United States Department of Energy, Office of Technical Information, PO Box 62, Oak Ridge, TN 37381, USA.

Universitätsbibliothek Hannover und Technische Informationsbibliothek UB/TIB, Welfengarten 1B, D-3000 Hannover, Germany.

University Microfilms International, UMI, 300 N. Zeeb Road, Ann Arbor, MI 48106–1346, USA.

Index